SHUTTLE
MISSION CONTROL

Flight Controller Stories and Photos, 1981–1992

MARIANNE J. DYSON

ISBN (print edition) 978-0-578-88252-9

SHUTTLE
MISSION CONTROL

by Marianne J. Dyson

Back in 1990 when this book was drafted, Mission Control consisted of the three-story, windowless section of Building 30 shown in this old photo of Johnson Space Center. The five-story addition, Building 30 South, was added in 1992 to the left of the three-story Control Center. NASA photo B30-682.

AUTHOR'S NOTE 2021

While sorting materials for donation to the Johnson Space Center Archive at UHCL, I came across a manuscript I wrote nearly thirty years ago, titled *Shuttle Mission Control*.

I was in a unique position to write that story. I had the privilege of serving as a flight controller for the first five Space Transportation System (STS) flights. I had joined NASA in 1979, coming in on the "ground floor" of the Shuttle Program. Six years and five flights later, I still loved my job, but with a husband who was also a flight controller, and no affordable childcare for unpredictable all-night shifts, I decided to find a more "normal" job until our children were older. Hernandez Engineering offered a perfect parttime position helping "customers" fly their experiments on the Shuttle and prepare for Space Station Freedom.

I was excited to be part of the commercial space era!

But the *Challenger* accident in 1986 changed everything. Commercial and military use of the Shuttle stopped. Freedom was subsumed into the Shuttle/Mir Program (announced in 1992) and morphed into the *International Space Station*. I pivoted from working in the space program to writing about it. In the decades since, I've won top awards for my children's books, coauthored books with Buzz Aldrin, and educated hundreds of thousands of people about space via writing and speaking. I documented my personal story in my memoir, *A Passion for Space: Adventures of a Pioneering Female NASA Flight Controller* (Springer, 2015). But the stories I collected from other controllers in 1992 remain mostly untold.

I drafted *Shuttle Mission Control* in 1990-92 when books were printed on paper and sold in bookstores. My agent and I collected a stack of rejection letters calling it too "narrowly focused" to sell enough copies to make publishers a profit. So the book languished in my closet all this time. Now, armed with decades of experience in publishing, no longer required to satisfy some gatekeeper's profit margin, and with the help of my writing community (special thanks to Tom and Lindsey!), I am publishing it myself.

Scanning, color-correcting, and cropping faded photos and slides proved difficult, as was reformatting/merging old text files (from 5.25" floppy disks!) and typing hand-written lists (sorry if some names are misspelled!) into a modern document. Tracking down interview subjects (sadly, at least one has died) allowed me to add (thanks, Paul!) some "where they are now" information. I also added updated Flight Control Room data and positions, replaced/supplemented drawings with photos, and enhanced the NASA provided generic captions with names of flight controllers pictured (thanks, Space Hipsters!).

To the lists of the first flight controllers to work each Shuttle "front room" position from STS-1 in April 1981 through STS-39 (the 40th flight, in April 1991), I added what data I could uncover about the first women/minorities using photos, contacts, and the Manned Spaceflight Operations Association (www.mannedspaceops.org) manning lists (thanks, Bill!). Flight Director Bob Castle supplied the list of flight controllers honored to hang the mission plaque after each Shuttle flight (thanks, Bob!).

The original photos/slides, references, and manuscript of *Shuttle Mission Control* will be donated to the NASA JSC Archive at UHCL in Houston. Proceeds from sales of this book will be donated to organizations and museums to help preserve more of the history of Mission Control.

Please report any errors, misspellings, or comments (keeping in mind that the interviews were conducted in 1992 and describe operations as they were at that time), via my website: www.MarianneDyson.com.

Thank you for taking the time to read about challenges faced and solutions found by flight controllers during the first decade of Shuttle operations. I hope you enjoy your visit to Shuttle Mission Control!

"The Mission Control Center must be considered among the most historically significant structures not just in this country, but in the entire world."

Stan Graves
Speaking for the Texas
Historical Commission

Built as a bomb shelter, Shuttle Mission Control was initially located in the three-story, windowless portion of building 30 at Johnson Space Center. Shuttle flights were controlled from the five-story addition starting in 1995. The building was named the Christopher Kraft Control Center in April 2011. NASA photo/James Blair July 2011.

MISSION CONTROL
ACHIEVEMENT THROUGH EXCELLENCE

Every time the Space Shuttle launched from the Kennedy Space Center (KSC) in Florida, flight controllers in Houston worked around the clock checking for clues of possible trouble onboard. With a two-billion-dollar vehicle and the lives of the crew at stake, failure was never an option.

The NASA Mission Control Center (MCC) at Johnson Space Center in Houston was first used for flight support of Gemini 4 in June 1965. Twenty years later, on December 24, 1985, the Secretary of the Interior designated the third-floor Flight Control Room as a National Historic Landmark. (Ref. 17)

Stan Graves, speaking for the Texas Historical Commission said, "Along with Launch Complex 39 at the Kennedy Space Center, the Mission Control Center most fully symbolizes this nation's space accomplishments which culminated in the successful Moon landing of July 1969. The Mission Control Center must be considered among the most historically significant structures not just in this country, but in the entire world." (Ref. 17)

The Mission Control Team Patch: the Greek letter sigma represents the total mission team, the sum efforts of individuals past, present and future. Skylab and Shuttle operations are highlighted on the patch while the border symbols represent the Mercury, Gemini, and Apollo programs. Res Gesta Per Excellentiam is the Mission Control motto: Achievement Through Excellence. (Ref. 26) Marianne Dyson photo 2021.

The original FCR-1 on the second floor of Building 30, shown here after the landing of STS-1 in 1981, was first used to support Gemini IV in 1965. Note, the lack of a mission patch on the H-14 column on the right side of the room distinguishes it from the nearly identical FCR-2 on the third floor. INCOs Ed Fendell and Lee Briscoe are in the forefront with Al Pennington behind them, and PAO Harold Stahl standing. NASA photo s81-30387

THE FLIGHT CONTROL ROOMS

Mission Control consists of a team of flight controllers that operate out of a flight control room, abbreviated FCR (pronounced "ficker"). Two nearly identical flight control rooms, one above the other in the three-story Building 30 at the Johnson Space Center (JSC) in Houston, Texas, supported all human spaceflight missions starting with in 1965. (Mercury and the first Gemini flights were controlled from Kennedy Space Center.) In 1992, a five-story wing called Building 30 South, added two more flight control rooms.

The original two FCRs, 1 on the second floor and 2 on the third floor, were called Mission Operations Control Rooms (MOCRs, pronounced moh-kers). Up until 1995, one FCR was used to support the upcoming or current flight while the other was used to train flight controllers for later flights with different trajectories, payloads, and procedures.

Rooms adjacent to the FCRs in both Building 30 and 30 South were used as Systems Support Rooms (SSRs) or Multi-purpose System Support Rooms (MPSR, pronounced "mip-sir"). SSRs were also referred to as "the back room" as opposed to the MOCR which was "the front room." Additional controllers were located around the world to support the control team and provide expertise on commercial, government, international and military payloads and experiments.

Public visitors to Johnson Space Center were not allowed in the viewing rooms "behind" the FCRs during mission support. Prior to the opening of Space Center Houston (in 1992), visitors were only allowed into the viewing room of the second floor FCR 1 during training. The third floor FCR 2 was off limits to the public.

But both of the original FCRs were seen on TV during flights. Many people didn't realize that these were two separate rooms because they looked so

Except for upgrades to the screens and computers behind the scenes, and new mission plaques added to the walls (note two on the H-14 column) FCR-1, shown here during STS-61-B in 1993, looked much the same as it had during Gemini and Apollo even ten years into the Shuttle Program. NASA photo STS061-s-098. *Flight Director Milt Heflin and Capcom Greg Harbaugh pictured.*

FCR-1 was gutted in 1995, converted to a Life Sciences Center and then extensively remodeled to become the International Space Station Mission Control, shown here in 2006. Note the column is no longer labeled H-14, and the upper portion of the wall is now rust colored. NASA photo jsc2006e43860.

much alike. A careful observer could tell FCR-1 and 2 apart by the mission patches/plaques on the walls, one for each mission controlled from that room. For example, on the right side (as seen from the back) in both rooms, there was a column labeled H-14. In FCR-1, there was no patch on that column until STS-49 in 1992 which sports a sailing ship, while the column in FCR-2 had sported the round patch from Apollo 16 in 1972 showing the red stripes of an American flag.

The mission plaques were hung on the wall of the FCR at the end of the last shift of each mission. The plaques were placed on the wall by individuals or teams that the Flight Directors determined had made the most significant contribution to the success of the flight. A table listing the flight controllers thus honored during the Space Shuttle Program is included in Appendix A.

FCR-2 was designated a National Historic Monument in 1985 because of its use for the Apollo 11 Moon landing. In celebration of the 50th anniversary of Apollo 11 in July 2019, the room and consoles were reconfigured to look as they did back in 1969, complete with ashtrays.

MOCR-1/Second Floor FCR-1

The first mission controlled from Houston was Gemini 4 in 1965. Geminis 4 through 12 (1965-66) were controlled from the second-floor room called MOCR or FCR 1. Apollo 1 (a launch pad fire during a test that killed the first Apollo crew in 1967), Apollo 5 (unmanned, 1968) and Apollo 7 (first manned Apollo, 1968), all three Skylab flights (1973); the Apollo-Soyuz Test Project (1975), and the first four Space Shuttle flights (1981-82) were controlled from the second floor. (Ref. 20 and photos.)

The second floor was also used to support later Space Shuttle flights in tandem with the third floor MOCR-2 (FCR-2) through 1992, and then all flights in 1993, 1994, and up through STS-71 in June 1995. The new White FCR (in Building 30 South) took over control of Shuttle flights starting with STS-70. (Yes, 71 flew before 70! Flights were assigned a number when placed on the manifest, but payloads were often delayed, scrambling the sequence. See Appendix A for the list of flights in order.) in July 1995. A table of which Shuttle flights were controlled from which FCR is included in Appendix B.

FCR-2, shown here at the conclusion of Apollo 11, was designated a National Historic Landmark in 1985. NASA Photo s69-40301.

In the summer of 1995, the second floor was gutted: all the consoles and the tiered flooring were removed. FCR-1 was converted to a Life Sciences Center at the end of the 90s and then, after extensive remodeling, began supporting the International Space Station (ISS) expeditions in 2006.

MOCR-2/Third Floor (Historic) FCR-2

The third floor, FCR 2, was used for (unmanned) Apollo 6, and Apollos 8 through 17. Also called MOCR 2, this control room on the third floor was specially outfitted to support classified flights. It was first used to support the Shuttle Program for STS-5 in November 1982.

FCR-2 continued to support Shuttle flights STS-6 through 9, STS-41-B, 41-C, 41-G; STS-51-C (first DoD flight), 51-D, 51-G, 51-I, and 51-J from 1983 up to and including the tragic last flight of *Challenger* and its crew on STS-51-L in January 1986. [Note, flights after 9 were numbered using fiscal year and launch site with "4" for FY 1984, etc. "1" for Kennedy Space Center, plus a letter A-L. NASA reverted to numbers after 51-L, starting with STS-26.]

The renovated FCR-2, complete with ashtrays and dial phones from 1969, opened to the public in July 2019 in time for the 50th anniversary of the Apollo 11 landing on the Moon. Marianne Dyson photo 2019.

As a result of STS-51L, the other Shuttles were grounded in 1987, and the "return-to-flight" Shuttle flight, STS-26, was controlled from the second floor. The next two flights, STS-27 and 28, had classified military payloads and were controlled from the third floor which offered secure communications. Four more Shuttle flights (STS-33, 36, 38, and 53) with military payloads were controlled from the third floor through STS-53 in December 1992.

FCR-2 was used in conjunction with FCR-1 for Shuttle flights from November 1982 through STS-53 in December 1992. This photo was taken during the deployment of Syncom IV (seen on screen) on STS-51D in April 1985. FDO Phillip J. Burley is in the center next to unidentified Air Force personnel seated at the Trajectory and GPO consoles which had not significantly changed since Apollo. NASA photo 51d-9099.

As noted earlier, because of its use during the historic Apollo 11 Moon landing, MOCR-2 was designated a National Historic Landmark in 1985. After its last use to support military payloads, the room was used only for special events and VIP tours. The viewing room behind the consoles was open for public tours through

Space Center Houston starting in the mid-1990s until it was closed for renovation in 2018. During renovation, the mission patches from Shuttle flights were removed from the walls as well as the console "tent" markers unique to the Shuttle Program. MOCR-2, as it was for Apollo 11, opened for public viewing just prior to the 50th anniversary in July 2019.

White FCR

In 1992, a five-story addition to Building 30, called 30 South, was completed. Building 30 South contained two new FCRs, designated Blue and White.

The White FCR became Shuttle Mission Control for STS-70 in 1995. Asst. Director of Mission Operations, Randy Stone (red tie), is at the MOD console during that first flight from the new FCR. NASA photo s95-16439.

The blue consoles in the White FCR, shown here in 2019, were removed and replaced with wooden desks and new computer hardware after the last Space Shuttle flight in 2011. Marianne Dyson photo, 2019.

The White FCR was Shuttle Mission Control starting with STS-70 in July 1995 through the last Space Shuttle flight, STS-135, in July 2011. The White FCR was then modernized (blue consoles replaced with wooden desks) and made available as a backup to FCR-1 for ISS operations and to support commercial crew flights such as the Boeing Starliner CST-100 flights to the ISS. (Ref. 43)

Blue FCR

The Blue FCR was originally called the Special Vehicles Operations Room (SVO) and was used to support the International Space Station assembly starting with STS-88 (which delivered the first American module, the Unity Node) in December 1998. Around-the-clock 24/7 operations in support of expedition crews began with Expedition 1 (with crew of three launched on a Russian Soyuz) in November 2000.

Control of ISS expeditions transitioned to the renovated FCR-1 in 2006. The Blue FCR was used for flight controller training and as a backup to FCR-1 for ISS operations through the end of the Shuttle

The Blue FCR, shown here during a Soyuz launch in 2000, was the original Space Station Mission Control, starting with STS-88 in Dec. 1998. From 2000 to 2006, the first 14 long duration expeditions were controlled 24/7 from this room. NASA photo jsc2000e27301.

Program. In 2012, the blue consoles were removed and replaced with wooden desks and modern hardware between to support use by commercial companies. The Blue FCR was Mission Control for the unmanned EFT-1 test of the Orion spacecraft in December 2014 and is slated for use by the Boeing CST-100 Starliner flights to the ISS.

The Bat Cave

The giant screens displaying spacecraft trajectory, live video, and flight data are an iconic feature of all Flight Control Rooms.

The original 10-by-20-foot screens in FCR-1 and 2 were not computer, television, or movie screens. They were half-inch thick glass lit from behind using slides. The projectors were in a room called the "bat cave" thus named because the room was very dark, and the floor was five feet lower than the rest of the floor. The bat cave had seven projectors to create its "special

After the Shuttle Program, the Blue FCR was upgraded with new hardware and wooden consoles which were used for the Orion EFT-1 test flight in December 2014. Marianne Dyson photo, 2014.

Shown here are employees of Bingswanger Glass, the company who made the original 10-by-20-foot screens, replacing the old with the new in 1990. The new screens were 800 pounds heavier but had better optical properties and could be easily washed. NASA Photo S9026138.

effects" like the giant world map familiar from the days when the first human entered space. (Ref. 20)

One background projector put a static world map on the screen using a *one-inch* slide. Then two "spotting" projectors placed symbols like a Space Shuttle or a satellite on the screen and moved them based on computer data.

Finally, four "scribing" projectors used diamond-tipped styli to scratch through a metal coating on glass slides, somewhat like a scratch-off lottery ticket. The scratches allowed light to pass through onto the screen. Several slides were superimposed to make one display, requiring careful alignment of the projectors. This system was designed before the invention of computer graphic displays, and it still worked in 1995!

But after twenty-five years of use and weighing over 1,200 pounds each, the original screens were nearly impossible to clean. So in 1989, the old half-inch thick glass screens were carefully removed using suction cups. "Everyone held their breath, waiting to see if the old glass would remain intact, or shatter," said one project engineer. Luckily, they came out in one piece. (Ref. 29)

Just above and to the left of the big screens was another landmark feature of Mission Control: the mission clock. Updated from crystal oscillators to an atomic cesium standard in the 1980's, this clock lost only one second every three thousand years. In the decades before cell phones, rumor had it that Mission Control visitors produced the greatest number of synchronized watches in the world. (Ref. 34)

Key Communications

Ever wonder what all those flashing buttons on the old consoles were for? Some of them were linked to data changes on the Space Shuttle, but many of them were simply telephone lines. Talking over ideas with other creative thinkers is often the best way to solve problems. Therefore, mission controllers needed to listen to the astronauts, the Flight Director, their fellow controllers, and people supporting them from offices, simulators, and other NASA or payload centers around the world.

The original communications system was a marvel of its time. A Gemini IV (1965) press release proudly stated the MCC had 52 million feet of wire—roughly the distance from Houston to the North Pole and back, or about 4,000 laps around the Indianapolis Speedway. (Ref. 23)

Originally, spacecraft had to be in a line of

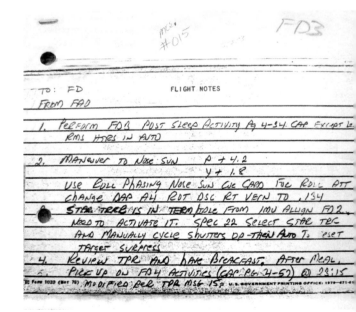

Flight controllers prepared flight notes, like this one describing the morning activities on the third flight day (FD3) of STS-3, for the Capcom to read to the crew during short ground passes. Copies were distributed to the team via p-tube and displayed on television monitors. Marianne Dyson photo, 2016.

sight with a ground station for the crew to send messages or data. Some of this data was collected and sent to Mission Control consoles via teletype circuits and undersea long lines. In 1983, the first of multiple Tracking Data Relay Satellites (TDRS) was deployed by STS-6. This network of satellites greatly increased communications between the spacecraft and the ground. Their descendants are still in use in 2021.

The old console phone buttons, each with its own little lamp, produced enough heat for mission controllers to warm their hands! These were replaced in 1991 by an energy-efficient digital phone system. (Ref. 20)

The other buttons and lights on Mission Control consoles were triggered by signals from the Orbiter. Examples included lights that came on at the start of a water dump or when a heater shut off. The status of Shuttle systems shown by these "indicator" lights plus information on the display screens was what the controllers talked about over all those phone lines.

Because seconds counted during short passes of spacecraft over ground sites, before any voice instructions were given to the crew, they were first written down on Flight Notes. The controller in charge of the instructions made multiple copies and sent them by pneumatic tube (P-tube) to support rooms, the controller in charge of the flight data file (FAO for Shuttle), the Flight Director, and the Capcom. Each note was discussed over the loops, and if approved by the Flight Director, read by the Capcom during the next acquisition of signal (AOS) with the crew.

Mission Operations Computers

The five original Mission Operations Computers (MOC, pronounced "mock") were considered "mainframe" computers. The computers took up the entire first floor of Building 30. Built in the 1960's by IBM, they each had only 128 kilobytes of memory and 880 kilobytes of storage (on magnetic tapes). It's not unusual for a mobile phone of the 2020's that a person can hold in their hand to have 256 gigabytes of memory, two million times more than what the old Mission Control could store. (Ref. 20)

The original five IBM computers took data, like spacecraft temperature, and ran it through programs to make graphs or show trends. These computers were accurate, but they were incredibly slow. Data from the long-duration Skylab flights came down faster than the computers could process it. Non-critical data from experiments overflowed onto disc drives and tapes: 4,500 tapes! It took almost <u>two years</u> to process and distribute the ten billion, one-hundred-seventy-eight million, five-hundred-seventy-one thousand, and four-hundred-twenty-eight parameters. (Ref. 20)

The Space Shuttle Program data requirements increased by an order of magnitude. On STS-9, the first Spacelab mission in 1983, approximately 720 BILLION bits of data were processed, and over 8,800 commands issued to the spacecraft by Mission Control. A network of over sixty-five workstations were needed to process this data and communications. (Ref. 20)

By 1994, Mission Control had increased its central processing capability to more than fifteen million instructions per second. (Ref. 20) The White and Blue FCRs were outfitted with more than 200 modular workstations linked together by 208 kilometers of fiber optic cable, the largest and fastest local area network in the world at the time. (Ref. 15) These workstations used commercially available software and

The Mission Operations Computers drove the status lights (upper grid) and displays like the one shown here on a console in 1980 during a simulation. The black and white display is a television screen, not a computer monitor. The handles on either side were to access the cathode ray tube (CRT) behind the screen. For the first Space Shuttle flights, there were 72 TV channels. This one is set on channel 2. NASA Photo S49(S) 360.

were interchangeable. Unlike the consoles in FCR1 and 2, the only difference between one station and the next is what data was requested by the flight controller. Therefore, if one console failed, a replacement was only a seat away.

Because human beings can only process a few things at a time, the computers of Mission Control were programmed not just to process data, but to sort out priority items that needed the attention of the human operator. The original consoles had display lights that were hardwired to turn on when a certain signal was received from a spacecraft. The modular consoles displayed those "lights" on a computer screen when spacecraft parameters met criteria or trends coded into software programs.

The Human Interface

Warning! A fuel cell has failed! While alarms sounded onboard the spacecraft, down in Mission Control, lights flashed on consoles and displays. Graphs showed trends that led to the problem or comparisons with data from previous flights. Using this information, mission controllers pinpointed the cause of failures and advised the crew what to do.

In the 1980s and early 90s, new equipment or software on the spacecraft meant new buttons had to be wired to consoles. Buttons used to track equipment no longer on the spacecraft were disconnected or moved like lamps from one room to another. All this "reconfiguration," took engineers weeks and sometimes months. Every change had to be carefully done to prevent false alarms or lack of real alarms during missions. By the mid-1990s, instead of physically moving lights, computers reprogramed the "wires" (the inputs) to them. Flight controllers choose the parameters to drive the lights on their console. With computers in charge of reconfiguration, the selected set didn't just change with the flight, but often many times during a flight.

Can you imagine a computer without a monitor? Outputs from computer "black boxes" of the 1960's were usually printouts of numbers or letters on paper, not on screens. But mission controllers

Paper copies were sent by and to controllers through a pneumatic tube system designed in the 1960's and used until 1995. The tube was sent by lifting the metal door in front of the stack of tubes, demonstrated here by INCO J. E. Conner in FCR-1 during STS-49. NASA photo S49(s)-362.

wanted their data displayed on a TV screen, not just as numbers, but with labels saying what each number or letter meant. So, they invented a Display and Control System to do this for them.

They took 35 mm slides to use as background like backdrops for a play. Placed between a slide and the inner face of a TV screen, was a set of stenciled letters or computer-drawn lines. The computer selected a slide and a set of stencils to match the data requested by a flight controller. Composite black and white video pictures were then sent to multiple flight controller consoles via various television channels. One hundred thirty-six television cameras and three hundred eighty-four receivers, the largest collection of television equipment in the world at the time, was needed to do this job. (Ref. 20)

NASA replaced this system in the 1970's with a Digital Television System. The new hardware replaced the slides with background text, and the old stencils with characters and lines which could be updated every second. For the Shuttle era, this system was in turn replaced with Display Generation Equipment, boosting the number of channels to 104. (Ref. 20)

Controllers could not easily record the data presented on their television screens. Video recording

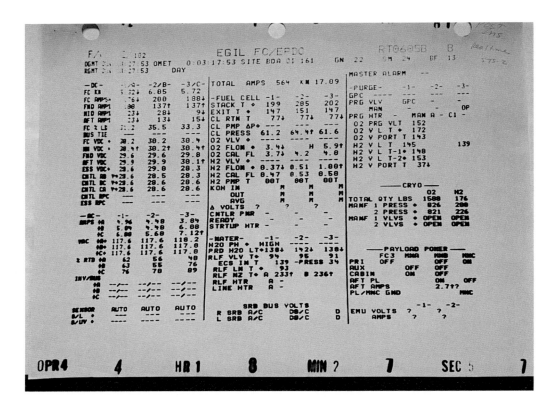

Controllers pushed a "print screen" button on their console to get a photo taken of displays and capture the data for review and postflight analysis. Shown here is a photo print copy of an electrical power and distribution and control (EPDC) display requested by the author during STS-2 when the fuel cell (FC) was failing. Marianne Dyson photo.

devices were not yet readily available and video cassettes to store them on were not invented until the 1970s. To get a paper copy of data displayed on a television screen In Mission Control, controllers took photographs of the displayed slides and data combined. The controller would push a button on a console to request a copy of what they saw on their TV channel. People in another room would tune in that same channel and take a photo of it with a 35 mm film camera—digital cameras were not yet available. This picture was then developed and printed on paper. It took about twenty seconds to dry, and then it was sent to the controller. This process allowed controllers to get "snapshot" records of their displays. (Ref. 23)

Snapshots and eventual paper copies were sent by and to controllers through a pneumatic tube system designed in the 1960's. The system used forced air to push cylinders up to two miles. Four-inch tubes connected over fifty stations on three floors of the building. The p-tube carriers travelled at about twenty-five feet/second. (Ref. 23)

There are no p-tubes in the modern Mission Control. With interactive computer screens to send and

P-tubes like the one shown were used to send paper copies of displays, flight notes, and other messages to and from flight controllers in different rooms. These were discussed "on the loop" with selected controllers using buttons seen under the phone dial behind the p-tube. Controllers headsets plugged into jacks that allowed them to listen to the crew, Flight Director, and other flight controllers simultaneously. To talk, a controller pressed a push-to-talk button on their headset and the phone button for the person (e.g. Flight Director was FD) which would then blink. Protocol was to say the position being called followed by the call sign of the position calling, such as "Flight, FAO." Marianne Dyson photo, 2014.

receive messages electronically, and printers connected to each console for paper copies, they weren't considered necessary.

The two FCRs with the big screens, consoles of buttons, multiple conversations, computers, and even the "thunking" of p-tubes were familiar parts of the old Mission Control. But even though the new FCRs look different, Mission Control remains a team of people who are adept at solving problems during spaceflight missions.

The Space Transportation System (STS) included the reusable winged Orbiter, the central orange external tank that held the fuel for the three main engines used only during launch, and the solid rocket boosters, one on either side, which provided thrust for the first two minutes of launch. Control of the STS switched from the Launch Control Center at Kennedy Space Center to Mission Control in Houston a few seconds after liftoff when the "Shuttle cleared the tower" or gantry visible to the lower left in this STS-5 image. NASA photo, STS-5.

In this photo taken in 1979, a NASA representative provides an overview of the workings of Shuttle Mission Control to a tour group seated in the viewing room overlooking the four rows of original consoles facing the Earth map in Flight Control Room 1 on the second floor of Building 30. NASA photo S79-93-951, cropped by Marianne Dyson.

THE POSITIONS

Depending on the mission phase, there were 18-22 active console positions in Shuttle Mission Control. Most of the positions required a college degree in math, science, or engineering. Some controllers were medical doctors and pilots, and the public affairs officer was usually a trained journalist with some technical background. In addition to technical skills, mission controllers needed to be good problem solvers, able to work on a team, and think fast under pressure.

Shuttle flight controllers either worked for aerospace companies or directly for the government as NASA employees. But civil servants and contractors alike followed the orders of the Flight Director (Flight) who made the final decisions in Mission Control. (The abbreviated position name, like FLIGHT, was the "call sign" of that controller and appeared on a brown or green placard on top of the console.)

In the historic Mission Control, the positions were assigned to consoles arranged auditorium-style in four raised rows facing the large screens. The row directly in front of the screens was called the <u>Trench</u>, being closest to floor level and also involving the most basic elements of flight control.

On the left end (facing the screens) of the trench row was the Ground Controller (GC) in charge of the Mission Control computers and ground network. Next in line was the Flight Dynamics (FDO, pronounced fi-doh)) and Trajectory (TRAJECTORY) officers who plotted the vehicle's course. The Guidance (GUIDANCE) Officer, later renamed Guidance and Procedures (GPO) officer and accompanied by the Rendezvous (RNDZ) Officer was in the center, making sure the Orbiter knew where it was and where it was going. The right side of the trench was the Data Processing System (DPS) Engineer who

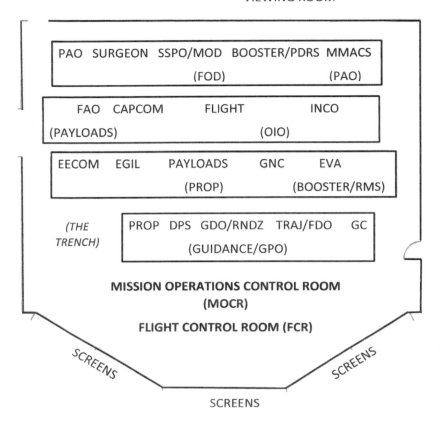

VIEWING ROOM

PAO SURGEON SSPO/MOD BOOSTER/PDRS MMACS
(FOD) (PAO)

FAO CAPCOM FLIGHT INCO
(PAYLOADS) (OIO)

EECOM EGIL PAYLOADS GNC EVA
(PROP) (BOOSTER/RMS)

(THE TRENCH)
PROP DPS GDO/RNDZ TRAJ/FDO GC
(GUIDANCE/GPO)

MISSION OPERATIONS CONTROL ROOM
(MOCR)

FLIGHT CONTROL ROOM (FCR)

SCREENS

SCREENS

SCREENS

The two original FCRs used for Space Shuttle Mission Control had four rows of consoles facing the viewing screens in front with the first row called The Trench. Positions merged, moved, and changed names as the program evolved. The console arrangement is shown as it was for a "generic" flight in 1992. Based on photographs and controller memories, some early names/placement are shown in parentheses. Diagram by Marianne Dyson, 2021.

was an expert on the Shuttle computer hardware and software. Propulsion (PROP) engineer in charge of the fuel systems and orbital maneuvering jets moved from the second row to the trench by STS-26.

The second row had four to six positions as the Shuttle Program evolved. For the early flights, the first console on the left was for the Booster (BOOSTER) engineer. Since BOOSTER oversaw the solid rockets and Shuttle main engines only used for ascent, the console was available for use by other controllers during the rest of the flight. Booster moved to the back row after STS-26.

The first four flights had no planned Extravehicular Activities (EVA), but some contingency cases called for spacewalks, so the EVA specialist and Crew Systems consoles were in the back room of the Flight Activities Officer. Likewise, the robotic arm, called the Remote Manipulator System (RMS), first flew on the second flight. The RMS console initially was in the back room of the Electrical Generation and Illumination (EGIL, pronounced "eagle") officer. For STS-2, RMS took over the Booster console after the launch phase. The arm was first used to deploy satellites on STS-5, so the position was renamed to Payload Deployment and Retrieval System (PDRS) specialist. This position moved to the back row with Booster for STS-26, but EVA remained in the second row.

EVAs often involved using the robotic arm to deploy and retrieve satellites and provide a platform for spacewalking astronauts, so the EVA and PDRS specialists sat side by side until STS-26. Starting with STS-26, PDRS moved with Booster to the back row to sit beside the new position in charge of crew and mechanical systems called Maintenance, Mechanical, Arm, and Crew Systems (MMACS). MMACS was the "handy man" type, in charge of just about any piece of loose equipment.

Next to EVA on the second row was the Guidance, Navigation, and Control (GNC) Systems engineer in charge of the sensors and controls that tell the Orbiter which way it is pointing (called its attitude) in space. The power system of the Orbiter belonged to the Electrical Generation and Illumination (EGIL, pronounced "eagle") Engineer. The EGIL position merged with the Environmental Engineer and Consumables Manager (EECOM, pronounced "ee-com") and RMS after the third flight. The two separated back out again beginning with STS-31. EECOM was always in charge of the air, water, heating and cooling systems.

The left console of the third row was occupied by the Integrated

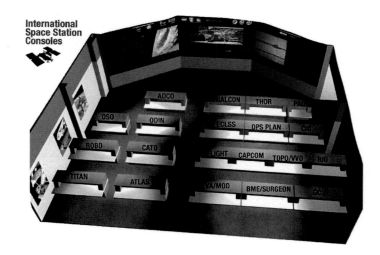

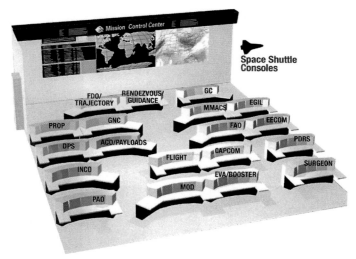

This chart shows the layout of flight control consoles in the ISS FCR 1 (starting in 2006) and White Shuttle (bottom) Flight Control Room from 1995 to the end of the Shuttle Program. NASA brochure, 2008.

Communications Officer (INCO, pronounced "in-coh") who got to remotely control the onboard television cameras and coordinate Orbiter communications systems. Next to INCO was the Operations Integration Officer (OIO) in charge of coordinating the details of the Flight Director's decisions. This position was phased out starting with STS-26.

In the center of the third row was the Flight Director's (FLIGHT) console. To become a Flight Director, a person must first have worked other console positions and qualify as a manager. Crew and vehicle safety were Flight's prime concern.

Next to the Flight Director was the Capsule Communicator (CAPCOM). For the Shuttle Program,

Capcom was always an astronaut. Capcom relayed all decisions by the Flight Director and all approved voice messages to the onboard crew. The Flight Activities Officer (FAO) got to tell the crew what to do and when, being responsible for the onboard schedule and changes to the vehicle's attitude and pointing.

For the early Shuttle flights, the payloads took a "back seat" to testing of the vehicle. So the Payloads (PAYLOADS) Officer's console was at the right end of the third row. As the importance and complexity of payloads increased, the position moved to the second row between GNC and EECOM where EGIL was originally. PAYLOADS' job was to make sure satellite and experiment needs were met which often involved coordinating with "customer" representatives located at other NASA or international space centers.

The "back" or "management" row, as it was sometimes called, contained a mixture of consoles. Starting with STS-26 (post-*Challenger*), the new Maintenance, Mechanical, Arm, and Crew Systems (MMACS) console took the far-left position with Booster/PDRS to their right.

In the center of the back row was an experienced Flight Director and senior manager, called the Flight or Mission Operations Director (FOD or MOD).

Next to the FOD position was a console for the head of the Space Shuttle Program Office (SSPO) and sometimes a representative of the Department of Defense (DoD).

Completing the back row were the NASA Flight Surgeon who was the crew's personal physician, and the Public Affairs Officer (PAO). During the early flights, the Public Affairs Officers (PAO) was on the far-left side, but later moved to the far-right side. PAO "translated" all the techno-talk into words the public could understand.

After every mission, it was traditional for the Flight Director to pick one person or position that most contributed to the success of the flight. That person or team got to hang the mission patch on the wall in the FCR. The following chapters provide a brief history of each position and a close look at the challenges faced by some of the flight controllers who were chosen to hang mission patches. (A list of all the controllers who hung mission plaques by flight is included in Appendix A.)

Though flight controllers had to master all the detailed aspects of their areas and systems of responsibility, this book only briefly describes and/or illustrates via diagrams and photos enough of the workings of those systems or elements of spaceflight to understand the stories told by the flight controllers. For more information about Space Shuttle systems, I recommend *To Orbit and Back Again: How the Space Shuttle Flew in Space* by Davide Sivolella (Springer, 2013).

As Eugene (Gene) F. Kranz, who served as Flight Director for Gemini, Apollo, and as MOD for Skylab, ASTP, and the Shuttle Program said, "The real reward [of working in Mission Control] is just the opportunity to be present in a room where you've got a magnificent team doing the marvelous things they do. It's like watching over the shoulders of a Michelangelo or a Da Vinci, being present when history is written." (Ref. 19)

GCs were responsible for the running of Mission Control in Building 30, the white, windowless 3-story building near the center of JSC, shown here in 1984. NASA photo s84-43-201.

GROUND CONTROLLER (GC)

If a flight controller had a problem with a display, headset, keyboard, computer program, or TV channel, they turned to the GC for help.

The GC managed the Mission Operations Computers, located on the first floor of building 30. At least two of these computers ran in parallel, with one of them designated as the Dynamic Standby Computer, ready to take over in case the primary computer failed. GC also coordinated ground tracking stations around the world and maintained the link to the Tracking and Data Relay Satellites (TDRS) through the Goddard Spaceflight Center in Greenbelt, Maryland.

The Ground Controller's job was not always done by one position. During Gemini and Apollo, the ground communications' functions were handled by a position called, "Network," and the displays and clocks were handled by the "Operations and Procedures Officer." As communications improved, and required less oversight, the network and facilities functions were combined.

The Ground Controller managed the Mission Operations Computers, which at the start of the Shuttle Program, consisted of five IBM 360/75 main frame computers shown here on the ground floor of Building 30. The computers converted telemetry data from the spacecraft into a form that could be displayed on the consoles. NASA photo, 1970.

The GC position, so named, came into being for STS-1, the first flight of the Shuttle Program. For the first ten years of the Shuttle Program, about 30 people served as GCs, all of them men. (See list at end of section.)

Ground Controller Honored

Shuttle flight controllers spent a great deal of time preparing for their console duties. Some of that time was spent in simulations—training that was just like the real thing. But much of a GC's time was spent in meetings. At one of these meetings, Ground Controller Larry Foy, who worked his way up through the ranks starting as a service technician at an Apollo tracking site in the Atlantic Ocean, was able to save a mission before it ever left the launch pad.

"During the STS-41 launch countdown (September 1990) the Shuttle was on the pad with the Ulysses Spacecraft in the payload bay," Foy explained. (Ulysses would eventually orbit the sun so scientists could learn more about it.) "The control center for Ulysses was at the Jet Propulsion Laboratory (JPL) in Pasadena, California. Through payload meetings, it became clear there was no

Ulysses spacecraft is shown here being mated with its Payload Assist Module at Kennedy Space Center in 1990. If the GC hadn't devised a way for JPL to check the software, an uncorrected error might have delayed or scrubbed the STS-41 launch. Instead, it was deployed and sent on its way to study the sun. NASA Photo KSC 90PC1172.

For his work in finding a way for the Jet Propulsion Lab to check the Ulysses computer, Ground Controller Larry Foy (on ladder) was given the honor of hanging the STS-41 plaque in Mission Control. INCO Joe Gibbs (hand on books) is at his console. Unidentified man and woman in forefront. NASA Photo S90-50282.

way for JPL to communicate with Ulysses to check everything was okay. I was able to devise a way for them to do this. The final agreement was reached about a week before flight.

"We anticipated the checks would not be needed, but during the check, JPL found an error in the software of the Ulysses computer and was able to correct it. If this problem had not been found early, the launch might have been delayed or scrubbed," Foy said. Launch delays and scrubs cost millions of dollars. "For my part in this," Foy said, "I had the honor of hanging the (STS-41) plaque." (Ref. 10)

Out of (Ground) Control

On STS-45 (March 1992), it was GC Larry Foy's quick action in real time that kept precious scientific data from being lost. The primary payload was the Atmospheric Laboratory for Applications and Science (ATLAS-1).

"Just as were entering the Zone of Exclusion (ZOE)," Foy said, "I got a call from the Network Director at Goddard Spaceflight Center." The ZOE was an area where the Orbiter was out of communications because it was not aligned with either the East or West Tracking Data Relay Satellite, usually over the Indian Ocean. "He said

During STS-45 in 1992, radiation caused a Tracking Data Relay Satellite (shown being deployed during STS-6 in 1983) to lose attitude control and tumble, cutting off communications between the Orbiter and Mission Control. Ground Controller Larry Foy routed communications to another satellite, restoring contact. NASA photo, 1983. (Ref. 24)

the TDRS West had taken a 'RAM (Random Access Memory) hit'. This particular satellite was the first one in the series and was not hardened against space radiation. This radiation, which is part of the normal environment of space, had caused loss of attitude control. It began tumbling.

"This caused us to lose contact with the Shuttle and crew," Foy explained. If the Orbiter were out of communications for more multiple orbits, the Shuttle's navigation data would deteriorate, requiring the crew to prepare for an emergency landing. As long as one TDRS kept working, the ground would be able to contact them before that happened. But there were other consequences to losing contact. "When we came out of the ZOE, we wouldn't get acquisition," Foy noted. "Also, the payload, ATLAS-1, whose instruments were studying the Earth's atmosphere, could not send its information to the scientists."

The data screens in Mission Control displayed static data while the operators waited tensely for Foy to reestablish contact. "Working with the Space Network Control Center at Goddard," Foy said, "we were

SHUTTLE GROUND CONTROL OFFICERS
STS-1 to STS-39

Name, first flight in MOCR

Charles M. Horstman, 1
George M. Egan, 1
James R. Brandenburg, 2
John E. Williams, 2
Don E. Halter, 3
Norman B. Talbot, 3
Robert R. Marriott, 4
David L. Holly, 4
John H. Wells, 5
Julius M. Conditt, 5
Charles R. Capps, 6
Wayne E. Murray, 6
C. Daniels, 41B
John Snyder, 41G
Mike Marsh, 41G
Edward H. Klein, 51D
W. A. Hopkins, 51D

CHALLENGER
Michael Marsh, (51L) 26
Robert Culbertson, 26
Alfred Davis, 26
Robert (Bob) Reynolds, 27
Larry Foy, 27
Victor Lucas, 29
Glenn Stromme, 30
Per Barsten, 28
Lynn Vernon, 33
Terry Quick, 41
Joseph Aquino, 41
Henry Allen, 41
Frank Stolarski, 35
Melissa Blizzard, (first woman), 43

Note, this list may be incomplete. No data was found for STS-51C, 51G, 51F, 51I, or 51J. (Ref. 48)

Atmospheric Laboratory for Applications and Science (ATLAS-1) flew on STS-45. The payload studied atmospheric chemistry, solar radiation, space plasma physics, and ultraviolet astronomy using 12 instruments from multiple countries. If Ground Controller Larry Foy hadn't reestablished communications with the Orbiter, much of the data Atlas-1 collected would have been permanently lost. NASA photo, 1992.

able to move the communications links to another satellite and regain contact with the Shuttle in about twenty minutes."

If Foy had not been able to move the links to another satellite, the Shuttle crew would have been reduced to intermittent communications at best. Normally, onboard recorders were played to the ground, called being "dumped," and then reused. If they couldn't be dumped, the recorders would fill up and additional data would have been lost forever. "It was another twenty hours before the original satellite was back in service," Foy said. (Ref. 10)

Foy continued as a GC until at least STS-69 in 1995. He worked for Allied Signal Technical Services through 1996, a contractor to NASA, and was part of the Integrated Planning Systems Office in the Mission Operations Directorate at that time. More recent information about him was not found in 2021.

FLIGHT DYNAMICS OFFICER
(FDO, Pronounced, "Fī-doh")

Dynamics means active—marked by continuous activity or change—a good description of the Flight Dynamics Officer's job. To reach space, vehicles must escape the gravitational pull of the Earth which requires reaching a speed of nearly eight miles per second. Pretty dynamic!

Brian D. Perry, who was a FDO for more than twenty-five Shuttle flights, summed up the FDO's job like this: "My position was in charge of figuring out how to get the best possible mission if something went wrong with the Shuttle's rocket engines. By best possible, we meant getting to an orbit that would allow completion of as many originally scheduled activities as possible.

"If the problem were serious enough that we couldn't continue to orbit, then my position determined where the Shuttle could safely land. Once in orbit, my position was in charge of knowing where the Shuttle was at all times, as well as where it would be during the rest of the flight. We analyzed and prepared orbit changes using the Shuttle's maneuvering engines and maintained plans for an emergency return to Earth."

1981 Shuttle Flight Dynamics and Guidance Officers in training (L to R): Phil Burley, Terry Burleson, Ed Gonzales, Rick Wray, Alan Keisner, Ted Dyson, Greg Oliver, and Brad Sweet. NASA photo.

To perform these tasks, the FDO had to know a bit about each of the other controllers' areas. "Although one didn't have to be a highly skilled orbital mechanic to do my job," Perry said, "an understanding of what's going on at the fundamental level greatly enriches one's ability to do it well. It was a real breakthrough for me during high school physics when an equation actually meant something I could relate to the real world."

Even with computers to sort priority data, there were too many critical events in the high-speed ascent and entry phases for one person to track. FDOs therefore had one of the most elaborate support rooms of all FCR positions. About thirty-five people supported the FDO during ascent, six during orbit, fifteen for rendezvous, and twenty for entry. These important people concentrated on the details of key data sets. They kept the FDO informed of problems that might lead to abort situations.

The FDO position has been around since the first flight controlled from Mission Control, Gemini 4. Unlike the dynamic job, the nickname "Fi-do" was been completely stable.

Only five people served as FDOs during Gemini, all of whom were also involved in the Apollo Program, one as Flight Director. Of the 25 people to be FDOs in the first 10 years of Shuttle operations, two started with Apollo. Jan McCoy served as the first female FDO on STS-30 in May 1989. She first served in the MOCR as Trajectory for flights 27 and 29. Her brother, Phil Burley, was also a Shuttle FDO. (Ref. 30)

Exceptional Service

During STS-51-F (July/August 1985), one of the Shuttle's three main engines shut down during ascent. Flight Dynamics Officer Brian D. Perry had to determine what action to take.

"We were near the point in the launch," Perry explained, "where if a single engine were to shut down, we could continue to a lower-than-normal orbit." It was already about ninety seconds too late for the crew to return to the launch site. They had only minutes to decide if they should make one orbit and return or continue to a lower-than-planned orbit.

The Flight Dynamics "back room" included multiple rows of consoles identical to the MOCR consoles but in a room with no big screens or windows. This snapshot includes Winds Thor (Ted) Dyson (blurred) and Bill Crimmell during prelaunch activities of STS-1. Marianne Dyson photo, April 1981.

"We have a complicated computer program which helps us determine what action to take. The analysis showed we needed to perform an "Abort-To-Orbit" to continue the mission," Perry said. This kind of abort places the Shuttle in a safe, but low, orbit. Perry said this was, "the first powered-flight abort in the US manned space program. It called into play all the skills for which the entire flight control team trains so hard and so long."

Then a second engine, "started showing the same indications of wanting to shut down as the first one had," Perry said. At least two of the three main engines were needed to reach orbit. "Losing a second engine would have been <u>real</u> bad." Many

When an engine shut down during launch, the crew of STS-51F (July 1985) used this rotary switch to select the first ever Abort-To-Orbit (ATO). NASA Photo 51F-18032.

FDO Brian Perry and Booster Jenny Howard (Stein) were given the honor of hanging the STS 51F mission plaque by lead Flight Director John Cox. Flight Controller Mike Collins appears in the background by Brian's shoulder. NASA Photo 51F (S) 272.

people held their breath while Mission Control studied the problem.

But before the second engine shut down, Booster Officer Jenny Howard (later, Stein) determined the engine was not the problem. Perry said, "A sensor on one of the engines failed so the computers thought the engine was violating a temperature limit. We told the onboard computers to ignore the sensor, which kept our two remaining engines going."

With this action taken, they were able to continue the Abort-To-Orbit rather than perform an emergency landing. An emergency landing, something never tried except in simulators, would have been dangerous and expensive. Even if there were no damage to the Orbiter, another multi-million-dollar launch would have had to be scheduled to make up the work this flight was to do. But, "the crew achieved a high enough orbit that the entire mission could be carried out," Perry said.

For their quick analysis which saved the mission, Brian Perry and Jenny Howard (Stein) shared the STS-51-F plaque hanging honors after the flight. They also were awarded, along with their Flight Director Cleon Lacefield, one of NASA's highest honors, the NASA Exceptional Service Medal. (Ref. 30)

Witness to Tragedy

Just six months later, Perry was on console for STS-51-L. This final launch of the Space Shuttle *Challenger* ended in disaster.

"At the time," Perry recalled, "FDO policy was to turn off the television feed prior to lift-off, in order not to be distracted from watching the critical data within the MCC. The indications on our displays were certainly not good, but there are lots of ways for data to go bad," Perry said. "It was more than a minute after the breakup before I found out what had really happened."

He conferred with the Range Safety Officer (RSO) at Kennedy Space Center and then had the grim duty of officially informing Flight Director Jay Greene, saying, "Flight, FDO, RSO reports the vehicle has exploded." (Ref. 55)

A commission headed by former Secretary of State William Rogers and including Astronauts Neil Armstrong and Sally Ride determined that a rubber O-ring in a joint between the segments of one of the solid rockets had not sealed properly and was the direct cause of the tragedy that took seven lives. Hot gas had escaped the seal, acted like a blow torch and melted part of the external fuel tank. The strut attaching the solid rocket to the tank broke, and the rocket nose swiveled and pierced the oxygen tank, causing an explosion. (Ref. 42)

On January 28, 1986, Flight Dynamics Officer Brian Perry had the unpleasant duty of informing the Mission Control Team that the Space Shuttle Challenger had exploded. The "Y" shaped plume was formed by the two solid rocket boosters which headed off in opposite directions. The orange fireball was caused by the ignition of propellants in the nose of the Orbiter. The crew of Rick Scobee, Michael Smith, El Onizuka, Judy Resnick, Ron McNair, Gregory Jarvis, and Christa McAuliffe died. NASA Photo.

After that day, Perry changed his policy to keep his monitor on through launch "in order to be aware as soon as possible to what is happening and be disciplined enough not to be distracted."

Perry will never forget that day he lost friends and co-workers. "It was one of those events that is just burned into one's mind." (Ref. 30)

A New Beginning

Perry was once again assigned as ascent FDO for the STS-26 return to flight. "For many of us," Perry said, "particularly those of us who worked 51-L, STS-26 was the 'first flight,' and we referred to it that way. We

Shown here is the picture-perfect launch of STS-26, the first flight after the Challenger tragedy. NASA Photo.

worked hard during the accident investigation, and we felt we had a whole new program -- one which was considerably better and safer and for which we had a deeper appreciation. I personally felt technically smarter as well as being somewhat wiser and more realistic about the dangers inherent in the manned spaceflight business." Perry received the Johnson Space Center Certificate of Commendation for his efforts to aid in a safe return to flight after the *Challenger* tragedy.

"I'm one of those guys who spent his childhood dreaming of one day working for NASA," Perry said. "I spent a lot of time reading and studying space stuff when most folks were out having a social life! Since many of the people who were astronauts and controllers during the Apollo days, and about whom I'd read, were still around when I came onboard, the 'living history' I stepped into was a dream come true." (Ref. 30)

Perry worked as a FDO until STS-53 in 1993. He left NASA sometime before June of 1996. No additional information about him was found in 2021.

SHUTTLE FDO
STS-1 to STS-39

Name, first flight in MOCR
Six TRAJ, listed below, were not also FDO.

Willis M. Bolt, 1
Jay H. Greene, 1
James E. l'Anson, 1
Craig Staresinich, 2
Ronald Epps, 2
Brad H. Sweet, 5
Gregory Thomas Oliver, 6
Brian L. Jones, 8
Edward Paul Gonzales, 8
Phillip J. Burley, 9
Ronald H. Cohen, 41B
Brian D. Perry, 41D
Chirold D. Epp, 51A
John Doug Rask, 51C
Wes Jones (TRAJ), 51C
William Ennis (TRAJ), 51C
Nicholas E. Combs, 51G
E. Mason Lancaster, 51F
Bruce Hilty, 61C
M. Sims (TRAJ), 61C
R. Sanchez (TRAJ), 61C

CHALLENGER
Timothy D. Brown, 26
Mark A. Haynes, 26
Kerry M. Soileau, 29
Matthew R. Abbott, 29
Jan McCoy, 30 (first woman)
Richard A. Theis, 28
Keith Fletcher, 33
K. D. Walyus (TRAJ), 33
Dan Adamo (TRAJ), 31
William Tracy, 38
Deborah Langan, 37

Note, this list may be incomplete. No data was found for 51I, 51J, 61B, and 27. (Ref. 48)

Guidance Officer Will Presley (second from left) at his console in the Mission Control "Trench" where he worked both Apollo and Shuttle flights. Fellow GPO Dennis Bentley is in the foreground. NASA Photo S26(S)091.

GUIDANCE AND GPO
Guidance Procedures Officer & Rendezvous

Our brains use our senses to tell us where we are, how fast we're moving, and in what direction we're headed. This position, speed, and direction information is called a state vector. The Space Shuttle computers used sensors to determine its state vector.

The Guidance Officer, later renamed Guidance Procedures Officer (GPO), compared what the Shuttle's computer said its state vector was with what the ground sensors said it should be. If the difference were large, the Shuttle "thought" it was somewhere it wasn't. The Guidance Officer's job was to correct deviations before they got large enough to be dangerous, possibly leading to loss of the vehicle during a rendezvous or entry. The procedures part of the job required a thorough knowledge of how to recover or respond to any failure which took the Shuttle off course.

The Rendezvous Procedures Officers shared a console with Guidance on flights that included a rendezvous—which was every flight to the Russian *Mir* or the International Space Station in the 1990s and 2000s. Rendezvous Procedures Officers were the air traffic controllers of space. When two vehicles going 17,000 miles/hour meet in orbit, even a small error in their speed or position can mean a deadly impact. With a limited supply of fuel in orbit, it was expensive, and sometimes impossible, to "fly around the pattern" and try again. The Rendezvous Procedures Officer's job was to plan the best approach, from above, below, in front, or behind the target, and to adjust the plan in case a problem developed.

Data All Squirrely

It is far better to prevent a problem than deal with it during a flight. Thanks to simulation training, many potential problems were uncovered and corrected prior to Shuttle missions. One fortunate example happened to Guidance Officer Thornton E. (Ted) Dyson while participating in a long-duration simulation

Four Shuttle flight controllers: L to R, Gail K. Weber, Marianne and Ted Dyson, Brian Perry pose in front of plane Ted rented to fly them to an airshow in 1981. Gail served as the first female Guidance on STS-7 in 1983, Marianne was the second female FAO in 1982, Ted was Guidance starting with STS-2 in 1981, and Perry was FDO starting with STS-41D in 1984. Marianne Dyson photo, 1981.

with the Entry Team prior to the first flight of Spacelab on STS-9 in 1983.

During the simulation, with the Entry Team on console in Mission Control and the crew in the simulator in Building 5 at Johnson Space Center, Dyson watched in alarm as the Orbiter's state vector "went all squirrely." Because of his experience working as Guidance on previous flights, he recognized the problem as a time tag issue. He explained, "The state vector has seven parts, three coordinates for position, velocity in three directions, and the time. The data was delivered in the downlink in two parts, called frames," he said. "Somehow, the two frames were being sent out of order, so that instead of two frames that go with the same time, I'd get one frame from a second ago, and the other from the current time. Knowing there were no jet firings in progress to change the velocity that quickly and seeing that the radar data showed the Orbiter moving steadily through space, I concluded the problem was with the ground telemetry processing computer (TPC)."

Dyson reported the time tag issue to Flight and recommended the standard fix for all computer issues: reboot, in this case, the telemetry processing computer. "This action fixed the problem, and the simulation continued," Dyson said. However, after the sim, he was charged with finding out <u>why</u> the TPC had generated bad data.

"I worked with a woman named Liz Browning, a contractor with IBM, who figured out the problem," he said. "Despite all the sims and flights that had already occurred, the TPC had never been left up and running for 24 hours straight. It'd always been cycled during loss of signal—but with the Tracking Data Relay Satellites, those times were not as long as they'd been for previous flights." Thus, the TPC had not been cycled for an extended time.

The problem impacted all the telemetry data, not just the state vector parameters. But temperatures and voltages didn't change that much in a second. "With the Shuttle moving at 25,000 feet/sec, the position values change rapidly," Dyson said. So he'd noticed the problem before the other console operators. "Once Ms. Browning discovered the cause, console procedures were changed to always reset the TPC prior to entry, insuring all the data was properly processed and interpreted during that dynamic time," Dyson said.

"After the flight, I wrote a memo to Ms. Browning's boss praising her work in solving this problem and improving the safety of our operations. I was surprised to get a call from her a few weeks later just gushing over how much she appreciated me writing that memo. Apparently, her management had little idea what she actually did for NASA. They were pleasantly surprised to find out that she'd played such a vital role. (Ref. 41)

Out-of-Plane

The Shuttle Guidance Officer was responsible for the navigational software in both the Orbiter's primary and backup computer programs. The primary and backup programs were similar, but not identical, because they were coded by two different groups. Like people who write different stories on the same topic, computer programmers write different code to solve the same problems. This prevented an error in one program from being automatically repeated in the other. At least one computer system, primary or backup, was mandatory at all times to operate the Orbiter.

Willard (Will) S. Presley, who worked the Guidance position from 1964 to 1990, was on console during STS-51-I (August 1985) when the Orbiter's primary and backup software programs disagreed.

"STS-51-I was a flight that used rendezvous targeting," Presley said. "One of the parameters in the targeting defines the inertial plane that you desire to achieve." An inertial plane can be thought of as the surface of a sheet of cardboard sliced through a model globe of Earth. For two spacecraft to rendezvous, or meet, in space, they need to be in the same plane. "On rendezvous flights, post MECO (Main Engine Cutoff), we have state vector limits for out-of-plane velocity errors."

"We had normal performance during the ascent and came nowhere near our powered flight update limits. We prepared for the OMS (Orbital Maneuvering System) 2 burn." This firing of the orbital engines, called a "burn," was needed to adjust the orbit. "Suddenly, I realized our out-of-plane velocity error did indeed exceed our limits," Presley said.

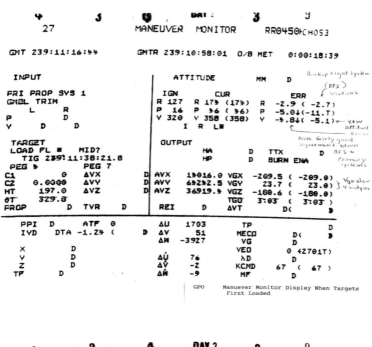

Compare these two copies of the Guidance console Maneuver Monitor display to see how Guidance Officer Will Presley discovered a software error that might have put the Orbiter into an incorrect orbit. (Ref. 31)

Somehow, the Orbiter's computer had not properly computed its speed and direction. Like a person playing pin-the-tail-on-the-donkey, the Orbiter needed an outside observer, Mission Control's computer, to "guide" it. This exchange of "guidance" information was called a state vector update.

"We decided to do a state vector update prior to the OMS maneuver," Presley said. "We did the update, and we also asked the crew to reload the targets (into the software). After they reloaded the targets, I looked at the burn solution, and I noticed there was not agreement between the primary and backup solutions. The backup had a body V-go that was larger than the primary."

The V-go was the velocity to go, in other words, the velocity change desired to reach the target orbit. The software written by the two groups had come up with different answers. The question was why?

"Everything else looked nominal," Presley remembered. Could the crew have typed in different numbers to the two systems? "I thought at the time that they hadn't loaded the proper OMS engine trims (fine adjustments to the direction the engine is pointing) in the backup. So we asked the crew to verify that. They did."

The crew had not made an error, so the Guidance Officer had to look elsewhere for an answer. But this was not easy in the case of the Backup Flight System (BFS). "We have quite a bit less insight, telemetry-wise, into the BFS," Presley admitted. "With the few things I had to look at, I couldn't tell anything that was causing the velocity solutions to be different." But he knew the backup numbers weren't right.

"Not knowing exactly what was wrong, we advised the crew prior to ignition, that if they had to engage (the backup system) during the burn, to steer just the in-plane velocities and disregard the out-of-plane velocity," Presley said. "If we had engaged (the BFS) and let the system fly its own solution, we would have had to schedule another maneuver in the rendezvous profile to get back into the plane we wanted." Like jumping up and down, changing planes uses energy. "It would have cut into the propellant reserves we had allotted for the rendezvous," Presley said.

Luckily, the primary system worked fine, and the backup system was not needed. "We did the burn, and everything was nominal," Presley said.

But something had caused the error, and Presley was determined to find out what. "We took hard copies (of our displays) and wrote down the steps we went through and submitted an anomaly (unexplained event) on the BFS. In real time, the flight control team kept an anomaly list for all failures and unexplained problems," Presley noted. "Each anomaly was worked and appropriate reactions defined and agreed to by upper management."

So, while Presley and the rest of the team went on with the flight, NASA management directed Rockwell, the company that built the Orbiter, to look at the anomaly. "Rockwell looked into the signature of what we had seen," Presley reported. "By the next day, they had checked their code and found an error. In rendezvous targeting, the parameter that defines the inertial plane is called IYD," he said.

IYD is a what is known as a unit vector, an arrow one unit long, that is normal to the desired plane. Normal in the mathematical sense means an arrow that is 90 degrees from the plane, like a pencil sticking straight up from a flat cardboard surface.

"Another parameter in the targeting is the nominal liftoff time," Presley explained. "If for some reason you don't liftoff at that particular time, there is logic in the onboard software that compensates by adjusting IYD for the radial progression (movement of the Earth) due to the slipped liftoff.

"Everything would have worked fine if we would have just loaded the targets once. After the targets were loaded the first time, it stored the new value over the original," Presley said. The flight had launched on time, but the backup computer thought it had slipped, so it changed the IYD parameter to account for the movement of the Earth. "When we loaded the targets the second time, it picked up the IYD that had already been corrected and did a double correction. The BFS was then targeted to a slightly different plane which led to the larger out-of-plane velocity solution we saw."

Usually, Presley said the Guidance position was most concerned with the numbers loaded, by the crew or ground team, into the equations and programs. "This was the only flight where we found a software error onboard," Presley said. As noted in Dyson's simulation case two years earlier, Presley said, "We go through so much software verification and simulation, that errors don't go undetected before flight. In this particular situation, we set the software in an unusual circumstance. All the sims and verification cases never required a state vector update after initial solutions.

"There's a lot of hours and inconvenient times you put in on simulations and training," Presley, who first worked the Guidance position during the Apollo Program, said of his many years of flight control work. "But the sheer excitement of watching a launch and knowing you made a contribution to a successful flight

SHUTTLE GUIDANCE
STS-1 to STS-39

Name, first flight in MOCR

Michael F. Collins, 1
Willard S. Presley, 1
J. T. Chapman, 1
Thornton E. (Ted) Dyson, 2
K. Alan Keisner, 6
Gail K. Weber, 7 (first woman)
Mason Lancaster, 8
Dave P. Kunkel, 51A
Joe Ken Patterson, 51G
Oscar Olszewski, 51F

CHALLENGER

Dennis M. Bentley, 26
D. Jeff Bertsch, 31
John M. Malarkey, 31
Andy Dougherty, 31
John Turner, 31
Matthew Glenn, 41
Glen Hillier, 37
Mark A. Thomas, 39
Chris K. Meyer, 39

Rendezvous
(first listed separately for STS-32)
James Oberg, 32
John M. Marlarkey, 32
Andy Dougherty, 32
Rick Hieb, 41C
Lynda Slifer, 48 (first woman)

Note, this list may be incomplete. No data was found for 51I & 51J (no release), 61B (no release), 27 (no data), 29 & 30 (not listed), 28 (no data), and 34 & 32 (not listed). (Ref. 48)

is pretty rewarding." (Ref. 31)

Presley retired from NASA in the mid-1990s and died in his chair with a drink by his side on July 16, 1997 at the age of 56. He's buried in Austin Memorial Park Cemetery in Austin, Texas. (Ref. 40)

Dyson left flight control after the *Challenger* accident and transitioned into information systems with a focus on safeguarding NASA's computer networks. He was the Johnson Space Center IT security manager when he retired from NASA in 2011. The owner of a Cessna 182, he flies as a volunteer pilot for the US Coast Guard Auxiliary.

James (Jim) Oberg, shown here in the PROP SSR for STS-1, was Rendezvous for STS-32. Photo courtesy Jim Oberg.

DPS ENGINEER
Data Processing System

The Gemini, Apollo, and Skylab programs all had onboard computers, but they were neither fast nor sophisticated. The crew relied on Mission Control's computers to process their data. The flight controllers interpreted it and recommended actions. As computers shrunk in size and weight (imagine trying to fly the 1960's Mission Control computer that took up the entire first floor of the building!), more data analysis could be done onboard.

By the Shuttle era, the onboard computers were a system of their own, requiring a dedicated flight control position to monitor them. In fact, the Space Shuttle, which was both an aircraft and a spaceship, was so complicated it couldn't fly without computers. Thus, the Data Processing System (DPS) Engineer position was created for STS-1 to oversee the operation of the Orbiter's five General Purpose Computers (GPCs). For the first flight only, there was also another front room position called Computer Command that was responsible for the interface between onboard computers and the ground.

The Data Processing System included, L to R, a General Purpose Computer, a Mass Memory Unit, keyboard, Display Unit, and Display Electronics Unit. Five computers were onboard. The three displays were in the cockpit between the pilot and copilot seats, and one in the aft flight deck. Programs that were only needed during certain flight phases were loaded when needed from mass memory units. IBM photo used with permission. (Ref. 13 & 16)

General Purpose Computer (GPC): The Shuttle Data Processing System Engineer was responsible for the five GPCs onboard the Space Shuttle. On the left was a "new" GPC that was half the size and had double the memory of an original GPC on the right. The "new" GPCs were installed in all the Orbiters in 1991. IBM photo used with permission. (Ref. 13 & 16)

IBM engineers tested Shuttle computers to verify they provided consistent and accurate results. DPS used this data to troubleshoot problems. IBM photo used with permission. (Ref. 13 & 16)

MARIANNE J. DYSON

Lit Up Like a Christmas Tree

DPS Officer Michael Darnell was faced with a real challenge during his first flight as a front room operator in April 1983. "During STS-6," Darnell said, "I was the DPS Officer on console for the first GPC failure during a mission."

The early Shuttle had four primary general purpose computers and one backup GPC onboard. As mentioned in the Guidance section, the backup computer was programmed by a different group of people to avoid a copied mistake failing all the computers at once. But on STS-6, it was one of the primary computers that failed.

"When this failure occurred," Darnell said, "my console lit up like a Christmas tree!"

All flight controllers received extensive training for contingencies such as this, and Darnell was no exception. "My team and I performed a set of malfunction procedures designed to cover such failures. These procedures were developed before flight to be used in the event a GPC failed."

Besides understanding what was wrong with the system and how to fix it, the flight team had to assess the recovery procedure's impact on the rest of the mission. For example, fixing the computer might take more time than the crew had available. Many of these impacts were thought through ahead of time and documented in a book. Darnell explained, "A set of Flight Rules governing the reconfiguration of the system was consulted while recovery of the failed GPC was in progress. Contingency procedures were [also] considered. The major one was the In-Flight Maintenance (IFM) procedure to replace the failed GPC."

But they didn't have to replace the failed GPC with an onboard spare. Instead, "the GPC was recovered, through the use of the malfunction procedures, and used during entry," Darnell said. "For my efforts, I got to

Michael Darnell's first mission as a Data Processing System Engineer also marked the first time a Space Shuttle computer failed. For his actions in recovering this critical piece of equipment, he was given the honor of hanging the STS-6 mission patch in Mission Control. NASA Photo S19-90055.

Elizabeth Cheshire, shown here with Randy Stone and Richard Truly during a sim for STS-1, was Computer Command which was a MOCR position for STS-1 in 1981. She was the first woman to be DPS Officer on STS-7 in 1983. NASA photo s81-3-9274.

hang the STS-6 mission plaque."

On future missions, Darnell continued to be challenged by system failures. "I was on console for the next GPC failure on STS-8 (August 1983)," he said. "It seems for every DPS failure (MDM's, MMU's, Input/Output failures), I was on console!"

However, people who successfully dealt with difficult challenges often earned, not only plaque-hanging honors, but leadership positions within NASA. As Darnell put it, "Working in the MCC is one of the most prestigious positions in the Mission Operations Directorate. From here, you can advance your career into management." (Ref. 8)

Darnell continued as a DPS through STS-26 in 1988. Then his prediction of moving into management came true. He became a mission integration manager, working flight and payload integration issues across NASA centers. When the Shuttle Program ended in 2011, he became the Johnson Space Center Engineering Directorate Space Act Agreement Coordinator, a position he still held in 2021.

SHUTTLE DPS
STS-1 to STS-39

Name, first flight in MOCR

Darrell E. Stamper, 1
Brock R. (Randy) Sone, 1
Kenneth W. Russell, 1
Emory Everett (Ernie) Smith, 2
Andrew F. Algate, 4
Michael Darnell, 6 (first Black)
Lizabeth A. Cheshire, 7 (first woman)
Gregory J. Harbaugh, 9
Ralph D. Monfort, 41D
Gerald (Gerry) W. Knori, 51B
Terry W. Keeler, 51F
T. M. Stevens, 51I

CHALLENGER
Gloria Araiza, 26
Mark D. Erminger, 27
Roberto Steven Galvez, 27
David D. Tee, 29
Burt F. Jackson, 34
James B. Hill, 41
Paul Tice, 35
Gary J. Sham, 39

(Ref 48)

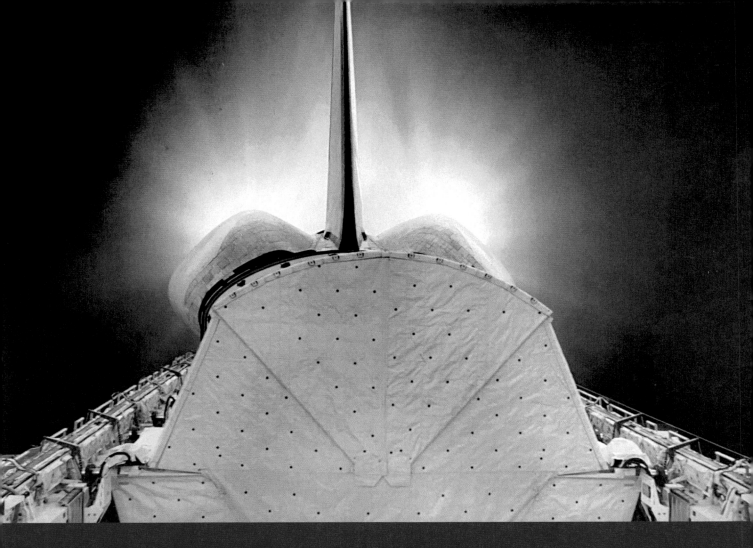

PROPULSION ENGINEER (PROP)

Running out of gas is never fun, but if you run out in space, you may never come home again. The Shuttle Propulsion Engineer's job was to make sure that never happened. (Note, the Space Shuttle used rocket fuel, not gasoline!)

PROP was in charge of all the orbit phase fuel users: two Orbital Maneuvering System (OMS, pronounced "ohms") engines used to circularize, raise or lower the orbit; 38 primary Reaction Control System (RCS) engines; and six vernier RCS thrusters used to maintain and adjust the orbit. (Ref. 28) The main engines were only used for ascent and were the responsibility of the Booster Engineer.

When the Orbital Maneuvering System engines (the two "humps" on either side of the tail) fire, the hot gases produced a glow shown here during an STS-5 experiment. Each OMS engine provided 26,688 Newtons, or six thousand pounds of thrust. The engines could be moved up, down, and sideways (gimbaled) to point the thrust in the best direction. Both engines were pointed inward after the deorbit burn so the Orbiter aft fuselage shielded them from the hot gases of entry. NASA Photo S-82-39799. (Ref. 28)

Great Planning

Propulsion Officers, like all flight controllers, had to be able to do more than one thing at a time, and do them all well. Propulsion Officer James (Jim) B. McDede, an aeronautical engineer, got to hang the STS-61B mission plaque in November 1985, "not for something that had gone wrong," McDede explained, "but for replanning activities.

"The 61B flight had a payload and experiment called the OEX/DAP," McDede said. "OEX stood for Orbiter Experiments and DAP for Digital Autopilot. Draper Labs had developed new flight software and new control laws that used the RCS thrusters in a different manner than what the Orbiter software normally used. They were assessing the propellant consumption.

"The concern was that because we were doing a flight test," McDede said, "the propellant consumption of the tests might get away from us, and we could impact the amount of propellant available for deorbit

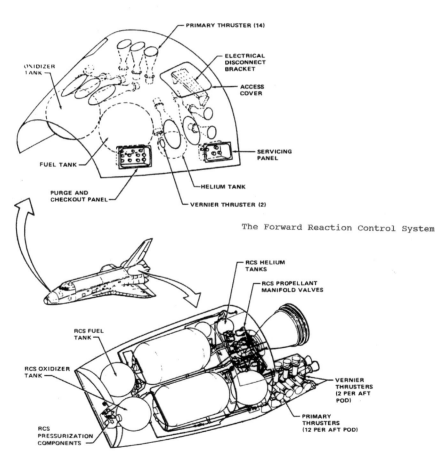

The Forward Reaction Control System

The Aft Reaction Control System

Forward RCS: There were fourteen primary and two vernier thrusters on the nose of the Orbiter. To optimize the center of gravity (balance point) of the Orbiter for entry, excess fuel in the forward tanks was "dumped," burned out of plane, using these engines. Like removing a penny from one end of a ruler, burning this fuel moved the balance point aft for better flight control. Vernier engines used less fuel and were quieter than the primaries. (Sound does not travel in a vacuum, but it does travel through solid objects in the hull to the crew cabin. Astronauts reported the loud banging kept them from sleeping.)

Aft RCS and OMS: There were twelve primary and two verniers on each of the two aft pods on the tail. Primary thrusters were 34 times more powerful than the verniers. Four of them could substitute for an OMS deorbit burn. Helium pushed the fuel or oxidizer out of their tanks to the firing chamber. The Propulsion Engineers used the helium pressure to tell how much fuel remained. (Ref. 28)

During STS-61B, the Orbiter had to fly in formation only thirty-five feet from the OEX/DAP payload, shown here being prepared for release by Astronaut Sherwood (Woody) Spring. The Propulsion Engineer's job was to make sure all this fancy flying did not use up fuel needed to come home. NASA Photo 61B-103-036.

and entry. Our job was to prevent that from happening and guarantee the propellant was available for deorbit and entry."

Propellant was carried in pods, one on each side of the tail fin (vertical stabilizer). The fuel was monomethyl hydrazine that explodes violently when exposed to oxygen. It was therefore kept separate from the oxidizer (something that adds oxygen, in this case nitrogen tetroxide), until a controlled burning in the form of an engine firing was desired.

"During the flight, there was a lot of replanning," McDede said. "On Flight Day 4, a crew constraint conflicted with a test the experimenter wanted. The flight team elected to move that test one day early into the current day's activities. Those activities included a thirty-five-foot station keeping operation with a target that was co-orbiting."

To station-keep means to fly in formation with another spacecraft, in this case, keeping the distance between them to only thirty-five feet. Objects in a lower orbit move faster than objects in higher orbits, so they constantly race ahead. Keeping the separation distance constant was a very difficult, and "gas-guzzling," thing to do.

"During those periods, the PROP console monitors the RCS propellant quantities, and in some cases, the OMS quantities through an interconnect mode." The two OMS engines were both on the tail of the Orbiter. "We can interconnect OMS and RCS systems and feed propellant to the RCS thrusters," McDede said. "It's a normal configuration, and we utilize it because the OMS [fuel] tanks are bigger." The RCS fuel, however, couldn't be used for the OMS because its fuel lines didn't support the extra flow rate needed by the larger engines.

"The addition of the OEX/DAP test to the end of Flight Day 4 required us to recompute the propellant allocation numbers," McDede said. "The only time to do that conflicted with the two hours of proximity operations [the station-keeping] underway, which were a high period of activity for the PROP console. We had to monitor the real time proximity operations while we were replanning the consumables as well as protect the prop levels for entry."

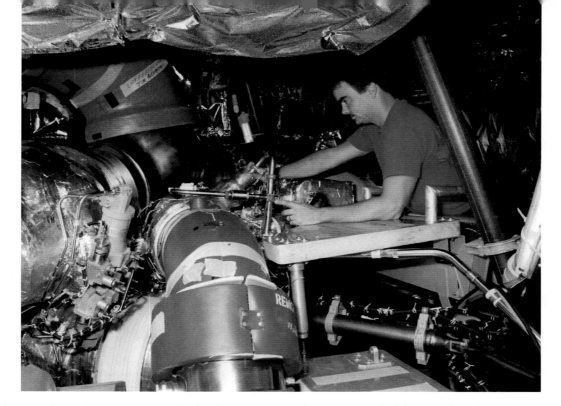

An engineer at Kennedy Space Center checks the connection of OMS engine fuel lines in the Orbiter. The Propulsion Engineers in Mission Control monitored the status of these lines during orbit. NASA KSC Photo 88PC976.

When faced with numerous difficult technical tasks, flight controllers must use every spare moment efficiently. For McDede, this meant using the precious few minutes when MCC was out of communications with the Orbiter to do the planning. "You get one or two of these outages over a two-hour period, and we were able to accomplish the replanning during those times. We completed the replanning about twenty minutes before the numbers were needed."

Like most controllers, McDede insisted on sharing his honors with his team. "That was a good example of the team really clicking and working well together," he said. "For doing that replanning during the middle of a high activity period for us, I got to hang the plaque." (Ref. 22)

From Childhood Interest to the Real Thing

Propulsion Officer Jim McDede, who left MCC to work on the Space Station Program, said, "I've always had an interest in the manned space program, going back to fifth and sixth grade where I was doing show-and-tell about the lunar landings to my classmates in science."

In 1971, his family vacationed in Florida so that the young McDede could see the Apollo 15 launch. "Four years later, as a senior in high school, I was back at Kennedy Space Center for the Apollo-Soyuz Test Project launch," McDede remembered. "Watching that launch cemented my career goals. I wanted to become an aeronautical engineer and go work for NASA."

He added, "Finally getting into the space business and participating in the Shuttle Program was a great

For their great planning work during STS-61B, Propulsion Engineers Jim McDede and Linda Hautzinger (later Ham) were given plaque-hanging honors by Flight Director John Cox. Appearing in the background are unidentified Air Force officer and FDO Brad H. Sweet (left of Cox), FDO Chirold D. Epp and Guidance Ted Dyson (right of Cox). NASA Photo S86-25330.

experience. For STS-1 [April 1981], I was a co-op student, then attending Embry-Riddle Aeronautical University, and worked for the PROP console on the Entry Team." After graduation, he was hired by NASA as a Propulsion Flight Controller. (Ref. 22)

McDede left the PROPconsole after STS-27 in 1988. He received a patent for a "Pre-integrated truss space station and method of assembly" in 1995 (filed in 1992). He worked in the Mechanical and IFM Branch of the Systems Division of the Mission Operations Directorate through 1998, and left NASA before the end of 2000.

SHUTTLE PROP
STS-1 to STS-39

Name, first flight in MOCR

Larry W. Strimple, 1
Gary E. Coen, 1
Glenn W. Watkins, 1
Ronald D. Dittemore, 2
N. Wayne Hale, Jr. 2
William H. Gerstenmaier, 4
Charles D. Young, 6
Richard N. Fitts, 7
Lonnie J. Schmitt, 41B
Anthony J. Ceccacci, 51A
Richard D. Jackson, 51D
Linda J. Hautzinger (Ham), 51B
(first woman)
J. H. Johnson, 51B
James B. McDede, 61B

CHALLENGER
Keith A. Chappell, 29
Sarah A. V. Kirby, 29
Karen E. Crawford (Jackson), 30
Matthew Berry, 31
Jeff Detroye, 31
William Powers, 35
Thomas Lazo, 37

(Ref. 48)

Linda Hautzinger (Ham) was the first woman to be PROP during STS-51B in 1985. Shown here discussing a problem with Capcom L. Blaine Hammond and Flight Directors Ron Dittemore (behind Blaine) and Lee Briscoe during a sim in 1987. She was also the first woman to serve as a Flight Director. NASA photo s87-4B553

Shuttle EVA training was done in the Weightless Environment Training Facility (WETF), shown here in 1989, until the Neutral Buoyancy Lab was completed in late 1995. The WETF was able to fit a mockup of the Space Shuttle payload bay but was too small to include space station modules. The WETF was filled with dirt and used for lunar and Mars training. NASA Photo S89-26715.

EXTRAVEHICULAR ACTIVITY (EVA) SPECIALIST

People have always wanted to fly. In space, they actually can. Because this kind of flying occurs outside of a vehicle, it is called, extravehicular activity, or EVA. The Shuttle EVA specialist was like a stage director, making sure the hero's suit and equipment worked, and that the actor safely followed the script.

Training Pays Off

EVA Specialist Bob Adams found STS-37 in April 1991 a good example of how being prepared for possible trouble could make all the difference. "When the Gamma Ray Observatory (GRO) was being deployed by the RMS, the high gain antenna got stuck," Adams recalled.

The antenna would relay data on gamma rays, a form of energy that cannot be viewed through Earth's atmosphere, to scientists on the ground. (Ref. 25) "No one in the Orbiter or on the ground knew what was wrong," Adams said. The mission of the Observatory's instruments—to witness the birth of elements, the death of stars, and discover clues about quasars, pulsars, and black holes—was at risk. But the EVA team was ready.

"Our group is unique," EVA Specialist, Bob Adams, explained. "We are not just flight controllers, but we also do a lot of hardware development and crew training. We put a lot of work into a lot of missions and then never got to use our stuff," Adams said. "It finally paid off on this mission."

In space, it took Shuttle astronauts about twenty minutes to put on a spacesuit. On Earth, it weighed two hundred and fifty pounds and could be a real challenge!

The first thing the astronaut put on was special long underwear. This underwear was not for keeping warm but keeping cool. Sort of like taking a shower while running, the undergarment surrounded the wearer with three hundred feet of water-filled tubes. The hands and feet were not covered, and thus often got sweaty and wrinkled like someone who's been in the bathtub too long.

Next on were the "pants" with boots attached. The hard upper torso was secured to the airlock wall. Holding onto the top of the pants, the astronaut squatted and slid up into the fiberglass "shirt," head popping out the top. The two halves were then connected with a waist ring. Last on were the gloves and helmet.

The backpack, or Primary Life Support System (PLSS, pronounced "pliss"), contained water storage tanks, oxygen bottles, a fan/separator/pump assembly, a filter, valves, sensors, communications, and a caution and warning system.

Air flowed down over the face and was collected at the wrists and ankles. If this air flow stopped, the crewmember's exhaled carbon dioxide would build up in the helmet, leading to unconsciousness and eventually death. A secondary oxygen pack provided an emergency thirty-minute supply to give astronauts time to reach the airlock, close the hatch, pressurize, and take off their helmet.

The collected air and water vapor passed through a filter that removed odors and carbon dioxide. Then it went through a fan to a cooling device called a sublimator that condensed the water out like dew. The removed water was added to storage for use in the underwear cooling tubes. More oxygen was then added to the cooled air, and the cycle repeated.

A drink bag was provided, but no food was included. No provision was made for bowel movements, but urine was collected into a bag for the men, and by a diaper for the women. These were disposed of in the Orbiter toilet after the EVA.

No crewmember ever vomited in a spacesuit, but if one had, it would have been very serious. If the wrist and ankle ducts got plugged, air flow would stop. Because of this danger, EVAs were rarely scheduled early in the flight when crewmembers were most likely to be nauseous from space adaptation syndrome.

A manned maneuvering unit (MMU) "jetpack" was used on three Shuttle missions, all in 1984. The MMU allowed astronauts to fly untethered in space. The MMU was mothballed after the *Challenger* disaster. [2021 update: a smaller unit called Simplified Aid for EVA Rescue (SAFER) is available for space station emergency use. (Ref. 48)]

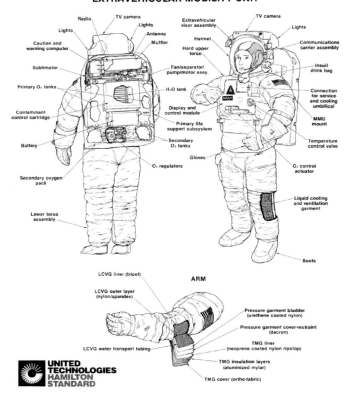

SHUTTLE EXTRAVEHICULAR MOBILITY UNIT

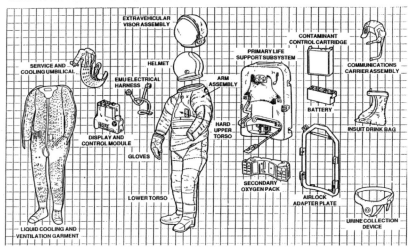

SPACE SUIT/LIFE SUPPORT SYSTEM OR EXTRAVEHICULAR MOBILITY UNIT

On STS-37, the Gamma Ray Observatory, shown here in painting by artist P.J. Weisgerber, required an unplanned EVA to fix the high gain antenna (bottom left) before it was deployed. (Ref. 25)

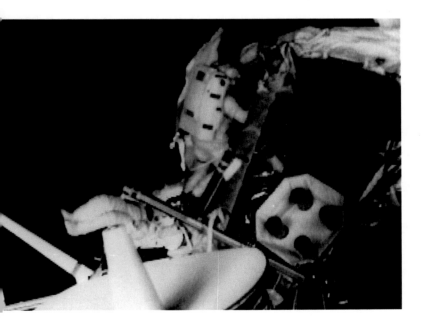

During an unplanned EVA on STS-37 in 1991, Astronauts Jerry Ross (top) and Jay Apt (boots showing) manually deployed the antenna (white cone at the bottom) of the Gamma Ray Observatory. NASA Photo S37 (S) 026.

The STS-37 crew had even more than the usual amount of EVA training because after the deployment of the Gamma Ray Observatory, they were scheduled for another day of EVA to test equipment for the space station.

"Almost every satellite we launch has some feature on it which could be used if something failed during the deployment," Adams explained. The GRO was the heaviest unmanned spacecraft NASA had ever launched, with each of the four instruments about the size of a subcompact car. (Ref. 25) Even though it was weightless in space, this huge mass still presented a danger and challenge to the crew. "But we had some procedures ready, some tools on board, and training completed that would allow the crew to go outside and deploy the antenna themselves," Adams said.

Low pressure was used in the Shuttle spacesuits because, like a balloon with less air, it was easier to bend in low pressure. If an astronaut were to go from high pressure to low suddenly, the air inside the body (balloon) tries to escape quickly. When this happens, bubbles of nitrogen in the blood can cause heart attacks and terrible pain in the joints—which is why this ailment is called the "bends." To prevent the bends, astronauts breathed pure oxygen to flush nitrogen from their blood. They could "pre-breathe" oxygen for several hours or lower the cabin pressure overnight and breathe oxygen for less time before going out. Because most women have a higher percentage of

EVA Specialist Bob Adams put his training to use helping the astronauts with an unplanned EVA during STS-37. For his work, he got to hang the mission plaque. Beside him are EVA team members, L to R, Charlie Armstrong, Oscar Koehler, Gerald E. Miller, Steve C. Doering. Behind two unidentified men in suits are Dave Simon and Bob Doremus. NASA Photo S37 (S) 091.

body fat than men, and this fat stores nitrogen well, most women must breathe oxygen longer than men to flush the nitrogen from their systems.

The STS-37 crew took a few hours to pre-breathe and suit up. Adams said, "Jerry Ross and Jay Apt went outside a few hours later."

This was the fourteenth EVA of the Shuttle Program, but that did not reduce the tension felt by the EVA Team. "It is pretty exciting to 'Go EVA,' and everyone kind of holds their breath the whole time," Adams said.

The first American to go EVA was Ed White on Gemini 4 in 1965. The first EVA of the Shuttle Program was planned for the fifth flight, but because of suit system failures, was postponed to STS-6 in April of 1983.

STS-37 was Ross's third flight and third EVA, but it was Apt's first flight and first EVA. Each EVA was unique, and no crewmember had ever done an EVA like this one before. "This was the first time we used an EVA crew to save a satellite we were trying to deploy," Adams said.

Although it may look easy to fly around in space, it is a lot of work. The physical strain, added to the limited amount of supplies the suit carries, restricted Shuttle EVAs to about six hours. Would six hours be enough for this unplanned EVA?

Yes! "They pulled on the antenna and unstuck it," Adams said. "This saved the $658 million-dollar satellite which would not have worked properly if the antenna were not deployed." The Gamma Ray Observatory was then deployed into its planned orbit and began collecting data on sources ten to fifty times fainter than those previously observed. (Ref. 25)

Susan Rainwater, shown here during STS-61 in December 1993, was the first woman to be an EVA. NASA photo STS061-s-097.

For his efforts, EVA Specialist Adams was given the honor of hanging the mission plaque in Mission Control. "I had put a lot of work into developing the EVA hardware and procedures," Adams said. "I'd tested the hardware to make sure it would work, trained the crewmembers and myself getting ready for this flight. The sense of accomplishment when everything comes together, and the mission is completed successfully is like finishing a race with everybody a winner." (Ref. 1)

Adams continued supporting EVA operations as a contractor with United Space Alliance (a NASA contractor until 2019) at least through 2006 (according to the JSC phone directory). The GRO collected data until the end of 1999 and was purposefully deorbited in June 2000. (Ref. 54)

SHUTTLE EVA
STS-1 to STS-39

Name, first flight in MOCR

James (Jim) A. McBride, 3
Barry E. Boswell, 4
Robert T. ᴬdams, 5
Charlie H. Armstrong, 6
(EMU Phase Support)
Terry Neal, 41B

CHALLENGER
Richard K. Fullerton, 27
James V. Thornton, 28
Wayne J. Wedlake, 34
Susan Rainwater, 31 (first woman)
Gerald E. Miller, 39

Was phase support for early flights.
First assigned for STS-41B. (Ref. 48)

The 1980s Space Shuttle cockpit had more displays and controls than even the most complicated fighter jet of its time. The GNC Systems Engineer was responsible for much of the equipment controlled by the switches shown here. NASA Photo S-80-29506.

In 1999, the Space Shuttle Atlantis was the first of the Orbiters to be outfitted with a "glass cockpit" (shown here in the fixed base simulator at JSC) which replaced the old flight control indicators and three green and black CRTs (in previous photo) with nine multicolored displays. NASA photo S99-01416.

GUIDANCE, NAVIGATION, AND CONTROLS (GNC) SYSTEMS ENGINEER

The Space Shuttle Orbiter had mechanical "eyes" and "ears" to tell the astronauts which way it was facing, if it was moving, and if so, how fast. These ears and eyes, the sensors, gyros, star trackers, and radios of the Orbiter, were the responsibility of the GNC Systems Engineer. Most of these navigational aids were controlled from the Orbiter cockpit with commands and data processed by computers.

Wasn't Supposed to be Possible

Guidance, Navigation, and Control systems engineer F. Edward (Eddie) Trlica, Jr. was on the GNC team honored for their work during STS-35 in December 1990. STS-35 was the first flight the GNC team was in charge of an instrument in the Orbiter's payload bay. Trlica said, "It was called the IPS for Instrument Pointing System, a system that controlled experiments mounted on it to point someplace other than where the Orbiter was pointing." Instruments like deep-space telescopes used the IPS to define exact locations of objects in the sky.

The ultraviolet telescopes of the ASTRO-1 payload, shown here in the bay of the Orbiter Columbia during STS-35 in 1990, relied on a computer system to find particular stars scientists wished to study. Thanks to the work of the GNC Systems Engineers, the mission was a complete success even though both the primary and backup onboard keyboards commanding the system failed. NASA Photo S35-13-008.

"STS-35, also called ASTRO-1, was the second flight of the IPS," Trlica explained, "but it was the first time we [GNCs] were responsible for that hardware. The reason they put it in our section was that we had a lot of familiarity with systems like gyros, accelerometers, and control loops."

Even though it had similar systems, the IPS was not a standard piece of Orbiter equipment. Trlica said, "It is kind of like Spacelab where they put a module in the payload bay."

"The Principal Investigators of each experiment prepared their instruments while our responsibility was for controlling the platform they were mounted on," Trlica said.

Controlling the platform proved to be an unexpected problem. "The IPS had two computers on it," Trlica explained. "One to control it, and one to control the experiments. The crew interfaced with those computers through two keyboards similar to ones on home computers."

Everything went fine for the first day of the flight. Then an unlikely pair of failures hit. "<u>Both</u> of the keyboards failed," Trlica said. "Because we could not point the IPS, the science mission was essentially lost."

However, Trlica, who started working for NASA before there was a Johnson Space Center, would not accept defeat. "When the first keyboard failed, we had one of our people start looking at what we might do if we had a second failure. For the first flight of IPS, they had a bunch of commands built preflight and stored in the Mission Operations Computer [in Mission Control, not onboard the Orbiter]. But after the Challenger accident, they decided to take those commands out for safety reasons," Trlica said. "The commands were needed only for that one particular flight, and people were afraid the commands might get sent [by mistake] even though there are a lot of protections against doing that." So Trlica knew there was a way to command the IPS from the ground, but first they had to recreate the commands that had been removed from the Mission Operations Computer.

"We knew what we needed to do," Trlica said. Using records from the previous flight, they came up with a way to work around the broken keyboards. "We knew what words we had to send to the IPS computer to do the job," Trlica said. "It was a matter of stringing words together in a particular order. We developed a bunch of commands to do the keystrokes from the ground and essentially took over the pointing of the IPS from our position."

But sending commands is not something GNCs regularly do. "Our position normally doesn't send very many commands at all," Trlica said. "We have a few commands to compensate the IMU [Inertial Measurement Unit, one of the Orbiter's movement sensors] when it drifts or something, but we normally send less than ten commands per flight. In four days on this flight, we sent about six hundred and fifty commands to the Orbiter!"

Not only was Trlica asked to do more commanding than usual, he was also required to work one and a half times his regular shift on console. "We normally have three teams [each working eight hours] supporting a flight around the clock," Trlica explained. "On this flight, one of our GNCs got sick on the second day. Because of the unique training for this mission, there were only three of us trained to do it, so the two [remaining] of us split the rest of the mission into twelve to thirteen-hour shifts." The onboard astronauts also worked twelve-hour shifts on this eight-and-a-half-day mission. "It worked real well," Trlica said, "But we were pretty tired by the end of the flight."

Even though the keyboards were failed onboard, the crew were still kept busy. "In several commands, we would load in a star, make the IPS move, then enable a joystick, like you might see on a computer game, that gave the crew manual control. The crew would take over from there and do the experiment. Then to do another one, the astronauts would have to maneuver the Orbiter in the general direction the [target] stars were. Once we told the IPS what the target number was, then the IPS would do the fine pointing."

Thanks to the work of the GNCs, the mission was a complete success. "In fact," Trlica said, "they got more than they thought they would be able to get. If we hadn't done what we did, they would have lost eighty percent of the data."

Trlica said it was a very satisfying flight for

STS-35 Mission Specialist Robert Parker manually points ASTRO-1's instruments using a toggle on the aft flight deck of the Shuttle. GNC Systems Engineer Ed Trlica had to manually transmit commands from the ground to save the science data. NASA Photo S35-10-011.unidentified men in suits are Dave Simon and Bob Doremus. NASA Photo S37 (S) 091.

STS-35 crewmember Sam Durrance uses a hand controller to maneuver the Orbiter in the general direction needed by the Instrument Pointing System (IPS). The GNC Systems Officer then moved the IPS by ground command. NASA Photo S53-12015.

GNC Linda P. Patterson, who worked a support room position for STS-1, is the second woman from the left in the STS-2 Ascent Team photo. The front row, L to R: Kathy Mintus (GNC SSR), Linda Patterson (GNC SSR), OIO Carolyn Blacknall, OIO William A. Middleton, ?Jan Wrather (not confirmed), INCO Al (Gap) Pennington, Marianne Dyson (FAO SSR), Ascent Phase Specialists Pearline Collector, Virginia Nestor (FAO SSR), Payloads Anngie Johnson. Behind Carolyn is Booster Jenny Howard (Stein). Behind Virginia is Terry Bantle (later Bagian, Paylaods SSR). NASA photo s81-39754.

him personally. "It's still a pioneering thing," he explained. "Even though we've been doing it a number of years, there are still a lot of unknowns." Like the dual failure of the IPS keyboards, Trlica said, "We haven't thought of everything that's going to happen or can happen. There's always something to learn." (Ref. 35)

Trlica continued working as a GNC at least through STS-76 (last data available) in 1996, and he was listed in the JSC phone book in the Guidance and Control Group through 2004.

SHUTTLE GNC
STS-1 to STS-39

Name, first flight in MOCR

Don J. Bourque, 1
Richard N. Fitts, 1
Harold J. Clancy, 1
Frank E. Trlica, Jr., 2
David W. Whittle, 4
Charles K. Alford, 4
James C. Adamson, 7
Harold Hardwick, 8
William L. Shelton, 41D
Linda Pritchard Patterson, 41D (first woman)

James Marshall Webb, 51D
Robert (Bob) E. Yackovetsky, 51B
Jeff W. Bantle, 51F
Robert Steven Galvez, 51G
Stephen J. Elsner, 61B

CHALLENGER
David E. Miller, 33
Heather M. Mitchell, 41
Kenneth M. Bain, 38
John P. Shannon, 38
Will Fenner, 35
Stanley Schaefer, 39

(Ref. 48)

The Payloads Officer was the liaison between the science and flight control teams. As payload operations became more complex, such as for Spacelab flight STS-51B in 1985 shown here, scientists monitored experiment operations from Marshall Spaceflight Center in Huntsville, Alabama. This facility evolved into the International Space Station Payloads Operations Control Center. NASA MSFC photo 8556032.

PAYLOADS OFFICER (PAYLOADS)

The early space program concentrated on understanding the spacecraft and systems needed to conquer the new frontier. Therefore, science took a back seat to the main objective of getting crews, and the equipment and procedures necessary, into space and back again safely. Science teams remained in the background during Apollo, monitoring their experiments and crew activities from support rooms and providing expert guidance and troubleshooting advice when called upon by the Flight Director or crew.

Science was central to the success of Skylab, leading to the formation of an Experiments console in the MOCR. This console was the first MOCR position staffed by a woman, Dr. Carolyn Huntoon, during Skylab 4.

The Experiments position was renamed to Payloads for the Space Shuttle Program. With different cargo and experiments on each Shuttle mission, the Payloads Officer's job focused on the interface of the experiment with Orbiter systems and the crew. Any malfunctions in power, cooling, or vehicle orientation, for example, might require changes to the experiments' operations or crew time requirements. Likewise, malfunctions on the payload "side" might impact Orbiter system use or crew time required for troubleshooting.

The first woman to staff a console in the Mission Operations Control Room was Dr. Carolyn Huntoon, the only woman in this photo of the Skylab 4 science team. No photos of her at the Experiments console are known to exist. NASA photo s73-32941. *(This is a photo of a photo taken in 2015—the original is no longer available.)*

As payloads became more complex and international, and communications systems more sophisticated, it was no longer feasible or necessary to have the entire science team physically located in a support room at JSC. NASA created a separate Payload Operations Control Center (POCC, pronounced "pock")

for the payload representatives at Marshal Spaceflight Center (MSFC) in Huntsville, Alabama. The Payloads Officer at JSC became the liaison for the POCC, relaying their requests, advice, and decisions to the Flight Director during real time operations.

Perhaps because interfacing with many different scientists from different fields and different cultures required exceptional "people" skills, the Payloads position attracted more women than other flight control positions. The first female flight controller to staff a MOCR console was Dr. Carolyn Huntoon who served as Experiments during Skylab 4 and later

STS-2 Payloads Officer Anngie Johnson shown here in her NASA office in the early 1990s, was the first black woman to staff a console in the Mission Operations Control Room. NASA Photo.

became the first female director of Johnson Space Center. The first black female flight controller, Anngie Johnson, served as Payloads Officer for STS-2. Even though they remained a minority, the Payloads Officer position continued to attract more women than any other flight control position. During the first ten years of Shuttle operations, other positions had one or two women, whereas over one third of Payload Officers were female. (Ref. 23)

An Alarming Situation

Michelle Brekke, Payloads Officer for STS-51G (June 1985), figured out that a satellite engine failure was not what it seemed. "I saw a rocket motor temperature go offscale high," she said. "It's kind of alarming to see a high temperature on a solid rocket motor. If it really were high, there would probably be an explosion."

For her quick diagnosis of a problem on STS-51G in 1985, Flight Director Larry Bourgeois asked Payloads Officer Michele Brekke to hang the mission plaque. NASA Photo S85-36671.

Since the payload with the solid rocket motor on it was still in the payload bay of the Orbiter, the consequences of such an explosion would have been catastrophic. "My immediate reaction was to kind of swallow my throat!" Brekke said.

"I began looking for some kind of confirming or denying clues. We had just lifted off, and everything was nominal regarding the Shuttle. There was no environmental effect that could have heated up the rocket [on the satellite in the payload bay]. There were other readings, so I figured if two were good, then this high [reading] was bad. It didn't take long to scan the display to another reading which was normal. Looking at that, I concluded it was an instrument (measurement) failure." (Ref. 4)

For her quick diagnosis and problem solving, Brekke was awarded the honor of hanging the STS-51G mission plaque.

Getting Out of a Jam

On STS-51I (August 1985), Payloads Officer Michele Brekke determined a course of action that saved a payload. "Right after we get on-orbit, some payloads require a health check to make sure they survived the launch [without damage from vibration or temperature extremes]. The Payload Assist Modules, PAMs, didn't meet the fault tolerance for inadvertent release or deploy, so we always had to have the sunshield open when we powered them [the satellites] up."

The sunshield protects satellites from getting too hot. "So we had opened the sunshield to do the health check," Brekke said, "but the crewman had maneuvered the RMS [Remote Manipulator System] elbow camera in such a way that when the sunshield opened, it pushed into the camera."

This happened during a time when Mission Control was out of communications with the Orbiter. Brekke explained, "On this flight, we were constrained on using the Ku-Band [communications relay through TDRS

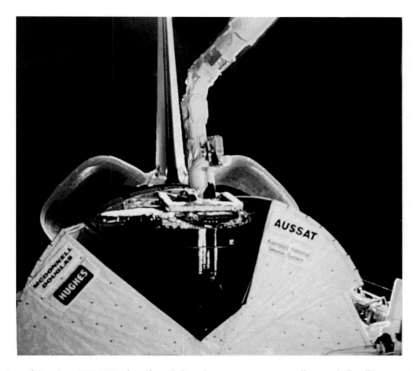

Oops! During STS 51I, the Shuttle's robot arm camera elbowed the "Pacman" sunshield protecting the AUSSAT satellite. The sunshield then got stuck on the satellite's antennae. Payloads Officer Michelle Brekke suggested the successful technique of using the robot arm to push the sunshield clear, allowing the satellite to be deployed. NASA Photo 51I-32005.

used for television coverage] because of the payloads. The crew comes up and says, 'Houston, we've got a problem here.' They explained it twice before we fully understood. Then we asked for downlink TV." Waiting for that TV took about thirty minutes, but Brekke said, "It seemed like an eternity."

"Once we got the TV, we could pretty much see what was happening. The sunshield got crumpled. It's pretty flimsy, like an umbrella, a very lightweight structure. It ended up getting deformed enough that it contacted one of the antennas on the payload in the cradle. It hung up on an antenna on the upper part of the satellite, and down on the bottom it was hung up on the elbow camera."

Brekke quickly assessed the options and came up with a solution. "I recommended we use the RMS [arm] to push the sunshield clear. Fortunately, the crew was able to unberth the arm. The next thing was to use the RMS to push the sunshield off." This action was successful.

Normally, the sunshield would have remained closed, protecting the satellite until it was time to deploy it. However, the sunshield was too damaged to close. "Sunshields are supposed to be reusable," Brekke said, "but that one was never used again. I think it's in a visitors' center somewhere."

Without a sunshield, the payload is vulnerable to the temperature extremes of the space environment. "We had to make sure we didn't violate any sun constraints," Brekke said. Therefore, the payload had to be deployed soon. Fortunately, "it turned out that particular satellite was the first one scheduled to go out," Brekke said. It was deployed without any problems a few orbits after the damaged sunshield was removed. (Ref. 4)

Wanted to be an Astronaut

Payloads Officer Michelle Brekke once dreamed of being the first female astronaut. "When I was sixteen,

Name, first flight in MOCR

John E. Hoover, 1
Tandy N. Bruce, 1
William J. Boone, 1
Anngienetta R. Johnson, 2 (first woman)
Janis Plesums, 2
Robert M. Kelso, 3
Jerry G. Renick, 4
Michael W. Hawes, 5
Jerry J. Conwell, 6
Linda M. Godwin, 6
James E. Duval, 6
Daniel D. Fennessy, 6
James Gauthier, 7
Debbie T. Pawkett, 7
Michele Brekke, 8
Glenn H. Cress, 9
William Middleton, 9
William Clint Burton, 9
Jay Apt, 51A
Michael Kenneth Fawcett, 51A
James Louis Clement, 51C
Sharon B. Castle, 51B
Colin Michael (Mike) Foale, 51I
J. R. Simons, 51J
Kathy V. Cannon, 61A

CHALLENGER
J. Mark Childress, 26
Ben L. Sellari, 26
D. E. Olsen, 27
Susan Lamb Creasy, 27
Neilan Eugene (Gene) Cook, 29
Nellie N. Carr, 29
David C. Schurr, 34
Tim E. Brown, 36
Jeffrey Hanley, 31
Cheryl Molnar-Boyd, 31
Mark A. Kirasich, 32
Debra Bulgher, 35
Roger Galpin, 35
Jean C. Costlow, 39
Sharon Conover, 39

Payloads Officer Michele Brekke, shown at the Flight Director's console during training, was selected as the first female Flight Director. She was in training when the Challenger disaster grounded all flights. During that downtime, considering that the Flight Director job required availability 24/7 and that she had small children at home, she decided to take a management assignment instead. (Linda Ham, a former PROP, became the first woman to serve as Flight Director in 1992.) NASA Photo S88-25474.

I watched the landing on the Moon and said, 'That's what I'm going to do.' I wrote to my Congressman and asked him what I had to study in college to be an astronaut. He wrote back saying NASA wasn't hiring lady astronauts, but if they were, I would need to get a jet pilot's license, four thousand hours of jet flying, and a master's degree in aerospace engineering."

Rather than putting aside her dream for a space career, Brekke came up with a plan. "I didn't want to be a jet pilot. But I figured NASA was going to need some non-pilot astronauts in the future. So, I went for the master's in aeronautical engineering and got it."

Her predictions about NASA's future needs were correct. "The very year I got my degree was the first year they announced for applications for mission specialists, and they *were* accepting women. I applied, but before I got my results, I was offered a job in the NASA crew training group to be an astronaut instructor. Two or three months later, I got

Note, this list may be incomplete. No data was found for STS-28. (Ref. 48)

Payloads Officer (later Flight Director) Rob Kelso (red hair) hosts the Shuttle's first student experimenter, Scott Thomas (foreground), during STS-5 in 1982. FAO Bill Holmberg is in the background. The Kangaroo (Roo) on top of the TV was the Payloads' Team mascot. NASA photo s82-39705.

my rejection [for mission specialist]."

Brekke worked as a crew training instructor for five years. "This was the best training possible for any Shuttle job," Brekke said. "Having to stand up in front of a classroom and teach a subject is great incentive to learn the subject yourself!"

Although her chances of being selected for the astronaut position increased with her experience at NASA, as time went on and Brekke married and had children, her desire to be an astronaut faded. Having worked with and trained astronauts for years, Brekke knew what that job would require. "Seeing how much they travel, and the off-hours commitment, like simulations until midnight, I couldn't see being an astronaut anymore. What was more important was I could be a full-time mommy and a full-time career woman with the job I had. It got to the point where I enjoyed my ground job so much, I never did apply again," Brekke said. (Ref. 4)

Brekke retired from NASA in 2019 after 37 years managing space missions. Still supporting customers of space, she went to work as a flight manager for Boeing supporting commercial space activities. (Ref. 53)

The Orbiter had three fuel cell power plants like the one shown here being re-installed in Discovery after being purged of fluids in 2011 in preparation for museum display. They were located under the cargo bay just behind the crew compartment. The fuel cells provided up to 2500 kilowatt-hours over a typical seven-day flight. NASA photo by Dimitri Gerondidakis.

ELECTRICAL GENERATION and ILLUMINATION ENGINEER (EGIL)

The EGIL (pronounced "eagle") had the power. The power system of the Orbiter, that is. The power system consisted of three power plants, or fuel cells. These giant batteries each weighed 202 pounds (92 kilograms) and were about the size of a four-drawer file cabinet (14x17x40 inches). Liquid oxygen and hydrogen were combined in these fuel cells to produce electricity for the Orbiter and pure drinking water for the crew. The EGIL was also in charge of the caution and warning system (smoke detectors, for example) and all the lights of the Orbiter. (Ref. 28)

Warning

Wee-oo, wee-oo! When alarms sounded on the Orbiter, and lights flashed red on consoles in Mission Control, something was seriously wrong.

Mark D. Fugitt was an EGIL who worked for Rockwell Space Operations Company (which was sold to Boeing in 1996). He recalled when a fellow EGIL, Robert E. Armstrong, Jr., was faced with those flashing red lights.

"During the STS-48 mission," Fugitt recalled, "a red warning light illuminated on the EGIL console. A "CLSD" appeared on the display indicating an oxygen valve had closed." A valve is like a door in a pipe that opens and closes. This valve led to one of the three Orbiter power plants (fuel cells). If oxygen couldn't get past the valve to the fuel cell, the fuel cell shut down. Without power, the Orbiter couldn't return to Earth.

The EGIL first checked to be sure the red light wasn't a mistake. "Errors sometimes occur during transmission, relay, and decoding of telemetry signals," Fugitt explained. But not this time. "The light confirmed the valve position status had been received by the control center. It was not a display error.

"The [same] message was annunciated onboard the vehicle a few seconds later," Fugitt recalled, "we knew it was a real problem versus a ground data problem."

During orbit periods, only two fuel cells were absolutely needed, but that did not make this problem any less scary. Like on any trip, when your vehicle starts having problems, it's time to get home before it gets worse, especially if your life depends on it. The EGIL knew unless the valve problem could be fixed, this mission would end soon.

"EGIL spoke to the Flight Director asking for immediate crew actions to prevent loss of the electrical buses," Fugitt said. Each fuel cell had its own bus, or power lines, to take power to equipment on the Orbiter. If the buses shut down, so did the equipment. However, one fuel cell could take over the bus of another. Since it seemed the fuel cell with the closed oxygen valve would soon quit generating power, the crew commanded the good fuel cells to take over the other one's bus and equipment.

"The crew performed the actions, and the situation was safed," Fugitt said. At least no equipment would suddenly lose power. Everyone watched and waited for the oxygen-starved fuel cell to die. Once it was declared dead, the Flight Director would have the team begin preparations to end the mission early like they had done when a fuel cell failed on STS-2. "For several extremely tense

Shuttle cryogenic oxygen tanks, seen here at KSC next to FAO Bob Nute in 1981, were spherical. Heaters extended through the center to increase the pressure as oxygen was depleted. Marianne Dyson photo, 1981.

minutes," Fugitt recalled, "EGIL had to prepare for loss of the fuel cell." But this was the fuel cell that wouldn't quit!

"The oxygen valve continued to indicate closed, but the fuel cell continued normal operation," Fugitt said. This was very strange because a fuel cell couldn't produce power without oxygen. (Liquid oxygen and liquid hydrogen were both needed to make electricity and water.) Therefore, despite the warning, the valve had to be open. "The EGIL officer was convinced the failure was not the valve closing, but a failure of the microswitch which provides valve status." Like a refrigerator light turning off instead of on when the door opens, "this was an instrumentation error."

The fuel cell was given back its share of the power load, and the mission continued. For his work uncovering the real problem, EGIL, Robert E. Armstrong, Jr., earned the privilege of hanging the STS-48 mission plaque. (Ref. 11)

EGILs Mark Fugitt (right) and Bill Gravett (left) at their office in Houston. Most EGILs are electrical engineers. All must complete hours of study and training to prepare them for the high-pressure job of console work. Marianne Dyson photo, 1995.

Shaking Under Pressure

All EGILs must earn their console "wings." Their ability to work under pressure is tested through intense and complex flight simulations until they prove they can handle it. As EGIL Mark Fugitt said, "When a problem presents itself, you must be able to focus on the task at hand, to remember the training provided, and handle the situation in a swift, accurate, and professional manner." Fugitt passed all the tests they could throw at him, and then had a chance to prove himself during his first Spacelab mission.

"STS-42 was a microgravity Spacelab mission," Fugitt said. "The microgravity environment is extremely stable with little relative motion [shaking] of the vehicle. Slight maneuvers were performed only once a day to allow for equal heating/cooling of the Shuttle exterior."

Fugitt carefully watched the effects this smooth ride was having on the liquid hydrogen and oxygen used to make the Orbiter's electrical power. The oxygen is kept at about minus 285 degrees F (-176 C), which is warm compared to the hydrogen at minus 420 degrees F (-251 C)! These very cold substances are called cryogenics, or cryo for short. They are stored in double-walled tanks like giant spherical thermos bottles.

Heaters inside the tanks warm the cryogenics just enough to make them expand out toward the fuel cell. When the cryogenics combine in the fuel cells, more heat is produced as well as electricity and water. If the extra heat is not removed, the cold cryogenics get mixed with the water vapor that normally "rains"

A veteran of Apollo, EGIL Bill Moon was a member of the Ascent Team for the first Space Shuttle flight, shown here in 1981. L to R, front row, unidentified, unidentified, GNC Richard N. Fitts, INCO Gap Pennington, Gary Coen, EGIL Bill Moon, EECOM Charlie Dumis, Payloads Tandy Bruce, Capcom Dan Brandenstein, Capcom Terry Hart, FAO Robert Nute, Phase Specialist Charles O. Lewis. Second row L to R: DPS Randy Stone, GC (?) Robert V. Grilli, Guidance Will Presley, EECOM Al Ong, Ronald Epps, unidentified, FDO Jay Greene, Booster John Kamman, Flight Neil Hutchinson, Command Joseph DeAtkine. NASA photo s81-31224.

out (for drinking water). Under these conditions, they don't produce as much electricity. Therefore, the EGIL must keep a careful watch on the temperatures of the cryogenics and the fuel cells.

"A side effect of operating heaters in cryogenic tanks is the stratification [separation] of cryo into layers of varying density," Fugitt said. [Density is the amount of mass in a certain size space. For example, a cube of metal is more dense than a cube of bread the same size.] "Since the heaters are axial, they extend through the center of the spherical tank. After a period of time, the fluid within the tank resembles an onion with each layer of skin having a different density. The center of the tank has the highest temperature and pressure." (High pressure means it has more energy and need to move. The center has the highest pressure because it is closest to the heater.)

"When a stratified tank is shaken, the mixing causes the pressure in the tank to drop as the layers mingle." (The high energy liquid in the center gets stirred in with the low energy liquid at the edges. Because there is more low energy stuff, the total pressure drops.) "Below a certain pressure, the liquid oxygen changes into a gas," Fugitt said. Like taking the lid off a pan of heated water, the liquid becomes steam as the pressure drops. But pure oxygen is much more dangerous than water vapor. "The heaters are prohibited from operating below this pressure because pockets of gas could form around the heater element allowing it to overheat and ignite [burst into flames]. Ignition inside an oxygen tank caused the Apollo 13 accident." (The Apollo 13 oxygen tank exploded, and the crew barely made it back to Earth alive.)

"On Flight Day 4, a [payload] bay North to South maneuver was performed. I had seen destratification [the layers mixing] effects two days earlier when the same maneuver was performed but was not prepared for the effects of this one," Fugitt remarked.

"On cue, the pressure in oxygen tanks one and two began to fall when the PROP Officer announced the maneuver complete. The pressure in tank one fell rapidly, but the heaters came on a few minutes later and the pressure rose. But the pressure in tank two continued to fall at a rapid rate even with all heater elements on."

"Quickly it passed the 730 psi [pounds per square inch] mark and headed for the 600's. When the pressure reached the 650 mark, I was quite concerned the tank would be lost.

"Messages were annunciated onboard and on the ground as the pressure dove. The crew and Flight Director asked for clarification of the situation. Discussion ensued about contingency plans," Fugitt said.

Were they in danger of an explosion? Could they finish the mission safely? Like a bunch of doctors who have done all they can do, the team watched to see if the tank would recover or continue to get worse. "Slowly, the pressure decay rate began to flatten out," Fugitt said. The tank was still getting worse, but not as quickly. "The lowest pressure reached was 590 psi," Fugitt reported. Finally though, the pressure stopped going down and began to climb again. With a great sense of relief, Fugitt watched as, "the heaters pumped the pressure back into the operating range."

SHUTTLE EGIL
STS-1 to 3 then STS-31 to STS-39

Name, first flight in MOCR

Jack Knight, 1
William J. Moon, 1
Paul M. Joyce, 1
Gary W. Johnson, 1
J. Milton Heflin, 2
William P. Gravett, 2
William (Bill) V. Bates, Jr., 3

CHALLENGER
Ray O. Miessler, 31
Robert E. Floyd, 31
Charles W. Dingell, 31
Robert E. Armstrong, 38
Mark D. Fugitt, 39
Unidentified first woman

Note, this list may be incomplete. EGIL was separated out for STS-1, 2, and 3, then combined with EECOM until STS-31. (Ref. 48)

Unlike Apollo 13, no explosion occurred. Fugitt recalled, "I took a deep breath, rubbed my aching eyes, and relaxed in my chair as the shift was drawing to a close. The Payloads Officer had their TV monitor switched to NASA Select [an in-house NASA broadcast transmitted to network TV stations], and I saw myself sitting in the MCC. At that moment I realized a tremendous feeling of accomplishment." All those hours of study and training had paid off. "It was an OH-WOW experience for sure," Fugitt said. "Then I got back to work." (Ref. 11)

Fugitt worked for Boeing for 21 years on the Space Shuttle and Space Station programs and then became a consultant. (Ref. 46)

The 23,400 lightweight, reusable tiles blanketing the surface of the Orbiter had to withstand temperatures up to 1260 degrees C (2300° F) and the 165-decibel sound blast of launch. Each tile was a 15-20 cm. (6-8 in.) "curved" rectangle cut to shape by a machine using a computer model of the Orbiter's surface. The tiles were installed, inspected, and replaced by workers (shown here in 1988) at Kennedy Space Center. (Ref. 12) NASA KSC Photo 88PC402.

ENVIRONMENTAL ENGINEER and CONSUMABLES MANAGER
EECOM, pronounced, "E-com"

We usually think of environment as those things needed to sustain life—such as air, water, plants, and animals. But the environment of space has none of these things. Environment means surroundings, and surrounding humans with space would mean instant death. To survive in space, humans must create an artificial environment. EECOM was responsible for the Orbiter's life support systems that provided this environment. These systems cleaned and circulated, heated and cooled the air and water and took care of waste disposal.

Apollo Experience

Jack Knight began his NASA career as an EECOM during the Apollo Moon program. He was promoted to chief of the Systems Division, manager of all the other EECOMS, and also many of the other flight controllers including the DPS, PROP, INCO, and Booster personnel. He was responsible for approving their promotions from support to primary positions and overseeing their training and certification.

Knight was one of many flight controllers who made the transition from Apollo to Shuttle, bringing his experience to the "new" program. Apollo gave him valuable lessons in dealing with the unexpected ways equipment reacts to the space environment.

"During one Apollo flight," Knight said, "just as we were completing cabin depressurization prior to an EVA [extravehicular activity], the Lunar Module coolant pump appeared to go awry. The primary data looked strange and did not correlate to the secondary data. We had to make a decision in fairly short order, or we were going to get behind on the EVA timeline."

The duration of EVAs was limited by the amount of oxygen and battery power each spacesuit had. "We watched for a few minutes, and the system seemed to recover," Knight recalled. "However, the automatic

The Shuttle EECOM was in charge of the infamous Waste Control System (WCS), or space toilet shown here in the middeck between the airlock (left) and exit hatch (right). Crewmembers used a seatbelt or tucked their toes under straps to keep from floating off the seat. Solid waste was sucked into the commode by a noisy vacuum system and then flung against the sides of a drum by rotating tines. Liquids were collected through a urinal assembly (white hose) that had adapters for men and women. A curtain (left) provided privacy. (Ref. 28) NASA Photo S86-26835.

pump switchover circuitry had selected the backup pump [indicating something might be wrong with the primary pump]. We had to decide if we wanted to stay [with the backup pump] or switch back."

Nothing like this had ever happened before, but Knight knew what to do. "I decided to switch back since the crew was going to be away from the Lunar Module. If another problem happened, I wanted the auto switchover available (it only worked in one direction)." If the backup failed, coolant would quit flowing.

"We later determined the problem to have been moisture on the sense line freezing as the pressure in the cabin fell," Knight said. "This caused a false indication and an automatic switchover [though nothing was wrong with the pump]."

The memory of another incident during Apollo, while serious at the time, brought a chuckle from Knight. "On Apollo 16 [April 1972], one of the EVA crewmen, Charlie Duke, decided he would do a flip on the lunar surface. He did not quite make it over and landed on his head! For a very looooong moment, we were all not breathing as we quickly looked over the telemetry from the suit and PLSS [Primary Life Support

EECOM Jack Knight will never forget being in Mission Control when Apollo 11 landed, July 20, 1969. "To be there and to participate when America landed on the Moon," Knight said, "was/is beyond words . . . it is still pure emotional ecstasy." Apollo 11 FDOs Dave Reed (white jacket) and Ed Pavelka (by flag) are in the foreground. NASA Photo S69-40299.

System]. Fortunately, nothing happened."

Memories like these are what working in the space program are all about, being there when history is made. "The opportunity to work for NASA, and in the Apollo program, was a dream come true for me," Knight said. (Ref. 18)

The First to Hang a Shuttle Plaque

EECOM Jack Knight was chosen by the Flight Directors to hang the mission plaque for the first Shuttle flight. "I suppose I was selected for the efforts I had been involved in throughout the Approach and Landing Test Program and in preparation leading up to STS-1," Knight said. "It was a real surprise, and I was extremely proud since nothing like that [flight] had ever been done before."

Having worked Apollo 13, which had an oxygen tank explode and barely made it back to Earth, Knight well understood the risks involved with the first Space Shuttle flight. "During early Apollo, we had followed the axiom of having to complete several successful unmanned flights before committing to manning a booster. However, on the STS program, we put two test pilots on the first flight of the full-up vehicle and went for it. There were a lot of crossed fingers on that one!"

The space environment was known, but like a new car on an old road, no one was sure how well the new systems would work. The systems EECOM oversaw had to protect the crew from +250° F (121° C) on the sun side, and -250° F (-156° C) in the shade. (Ref. 12)

Keeping cool was harder than keeping warm inside the Orbiter. Objects absorb heat quickly, like in a microwave. But it takes a long time to lose heat, as you know if you've ever waited for popsicles to freeze. The Orbiter was heated by the sun (sixty minutes of every ninety-minute orbit) from the outside and heated by equipment and people from the inside. So the Orbiter cooling systems had their work cut out for them, and so did the EECOMs in Mission Control.

"After we got on-orbit and got the payload bay doors open, John Young [Commander] reported missing tiles on the OMS [Orbital Maneuvering System engine] pods," Knight said. These pods bulge out on either side of the Orbiter's tail. Tiles on these pods and on the bottom of the Orbiter, like foam around a cold drink can, keep the Orbiter cool inside. "That really got our attention because we had no insights about what was going on on the bottom of the vehicle." The fear was that if too many tiles were missing on the underside, the vehicle would burn up during entry as the flight control systems overheated and failed.

On future fights, a camera on the robotic arm was used to survey the underside of the Orbiter, but this option was not available on the first mission. "There was a rumor that the USAF ground cameras were used to try to get some insight into the status of the bottom tiles, but the data was classified," Knight said. "Perhaps the Flight Director knew, but I did not."

Not knowing if the tiles were gone or not, Knight made sure the Orbiter cooling system was set to give maximum help. The cooling system had three parts: air, water, and freon. The crew and cabin equipment were cooled by fans blowing air. The air was cooled by water which was in turn cooled by freon. (Freon is toxic under some conditions, so it was not allowed to flow through the cabin and cool the air directly.) The freon released the heat it had absorbed into space through the payload bay door radiator panels. When the doors were closed, water from the fuel cells sprayed the freon loops to cool them. (This method was not normally used during orbit because the amount of water available was limited.) Once the Orbiter was back in the warm air of Earth, spraying water didn't provide enough cooling, so cold ammonia was used to cool the freon. (Ref. 28)

These systems checked out fine during orbit, but if a lot of tiles were missing, everyone wondered if they would be enough to keep the crew from being cooked during entry. Knight, and the rest of the flight control team, had to wait a few tense days to find out. "I was on the Entry Team and still vividly remember entry day," Knight said. For the first few flights, there were no relay satellites in orbit. This meant that during the critical heating period, Mission Control was out of touch with the crew. Would they burn up?

EECOM Jack Knight, second from right, was on console when pictures showed thermal tiles missing (see next photo) from the Orbiter Columbia on STS-1. It wasn't known at the time if the vehicle could survive entry heating without those tiles. NASA Photo S81-32526.

Soon after the payload bay doors opened on STS-1, this photo showed flight controllers that thermal tiles had fallen off during launch, exposing the surface underneath (dark squares on the white OMS pod) to the heat of entry. Unknown at the time was whether or not tiles had also fallen off the belly of the Orbiter. NASA photo sts001-008-0289, 1981.

EECOM Charlie Dumis is seen here during a countdown hold for the first Space Shuttle flight. Marianne Dyson photo, 1981.

SHUTTLE EECOM
STS-1 to STS-39

Name, first flight in MOCR

Jack Knight, 1
Charlie L. Dumis, 1
Jimmy Steve McLendon, 1
Albert Ong, 2
William (Bill) Bates, Jr., 2
J. Milton Heflin, 4
Paul M. Joyce, 4
Charles T. Holliman, 5
R. John Rector, 5
Jerry D. Pfleeger, 6
Mark D. Louis, 41B
Barbara N. Pearson, 41G (first woman)
William J. Moon, 51A
Joe A. Carretto 51C
Gary B. Evans, 51F
Dave G. Herbek, 61B

CHALLENGER
Charles W. Dingell, 26
Leonard J. Riche, 27
Ray O. Miessler, 29
Robert E. Armstrong, 28
Peter Cerna, 31
Quinn L. Carelock, 41
Daniel Molina, 37

(Ref. 48)

"When they came out of [communications] blackout after the deorbit burn," Knight said, "all our systems were purring along just like they were designed to. We all saw the image of the Orbiter on that marvelous long-range camera. What a sight!"

Mission Control erupted in cheers. As Knight said, "They were at eighty thousand feet or so and had come through the high heating region intact." Some tiles were missing, but enough remained that the cooling system had been able to do the job.

"From there on down," Knight said, "it followed the predicted profile and eventually came through the twenty-thousand-foot point looking almost exactly like the ALT [Approach and Landing Tests] profiles.

"They flared out on final, dropped the gear, and gently touched down at Edwards Air Force Base, coming to a stop about forty-five seconds later." Mission Control cheered again. Eight years after the last Apollo, America was back in space with a new reusable vehicle. Knight said, "Chris Kraft [Johnson Space Center Director]'s words still pertain, 'We just became infinitely smarter.'"

Shortly after that, Knight proudly hung the STS-1 mission patch on the wall of the second floor Flight Control Room. "I have always been a reader of science fiction, and I think NASA and the space program are the real hope and dream for the future of this country and the world." (Ref. 18)

Jack Knight worked as an EGIL again on STS-2, then as RMU on STS-3 and 4 before being promoted to management. As chief of the Systems Division of the Mission Operations Directorate, Knight watched over the flight teams from the Spacecraft Analysis (SPAN) room up through STS-30 in 1989. Knight was made Chief Engineer in MOD and was chief of the Advanced Operations and Development Division when he retired in 2006.

During STS-5, Astronaut Joe Allen used one of the portable television cameras while another crewman took this picture. The Shuttle INCO could remotely control these cameras as well as the ones in the payload bay from the ground. Video images were viewed onboard and transmitted to Earth or recorded on tape for later playback. NASA Photo STS5-38-966 or s82-39790.

INTEGRATED COMMUNICATIONS OFFICER (INCO)

Shuttle mission controllers often talked about their "telemetry," data transmitted from the space vehicle to their consoles; and their "uplink" which were commands sent to the Orbiter systems. The Integrated Communications Officer (INCO, pronounced "In-coh") was responsible for the telemetry and uplink which passed through the onboard communications system.

The onboard system included transmitters and receivers, signal processors, data recording and playback systems, television and voice systems. If another flight controller needed to reset, correct, or provide new data to the onboard systems, INCO checked and sent the commands. INCO was also the only one in Mission Control who could remotely operate equipment like cameras onboard.

Former INCO Bob Castle, who later became a Flight Director, said, "INCO is a great job. It's the only position in Mission Control where you actually control the systems you monitor. The [other] flight controllers suggest what is to be done, but the crew actually do it. For the INCO, much of the system is run from the ground by sending commands." (Ref. 6)

A Challenging Amount of Data

The Flight Director honors a person or console team who especially contributed to the mission's success by choosing them to hang the mission plaque on the wall in the Flight Control Room. Robert (Bob) Castle, INCO for STS-9, was so chosen for that flight.

"The first Spacelab mission, STS-9 [November 1983], was a very challenging mission for an INCO," Castle explained. "The Spacelab module added an entirely new set of communication hardware. There were new voice systems, TV systems, and other data systems. The entire mission depended on sending the data back to the ground via communication links."

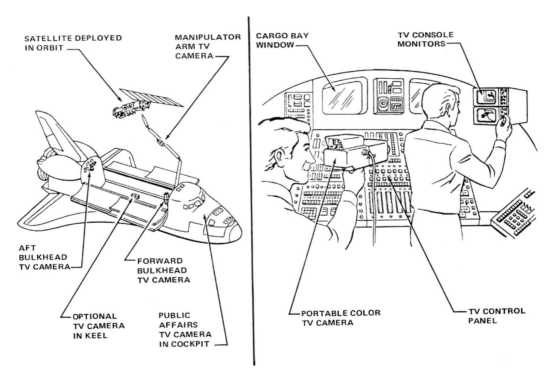

Orbiter closed-circuit television system.

The Shuttle closed circuit onboard television system could receive inputs from up to ten cameras. Portable cameras were available as well as cameras mounted on the forward and aft bulkheads of the payload bay and on the remote manipulator system (the arm). (Ref. 28)

But there was one big problem. Castle said, "The Tracking and Data Relay Satellite [there was only one at that time] had suffered a failure just before flight." This satellite relays communications from the Orbiter when it is on the other side of the Earth from Mission Control. Without it, all that remained were a few ground stations that could "talk" to the Orbiter a few minutes each while it passed overhead. This was like having a brand-new hi-definition television and finding out there was only one show per day and it was in black and white!

The Orbiter had recorders to capture some of the data collected while it was out of touch with Mission Control. But as noted earlier, like a cup of water, these recorders could only hold so much data. They had to be dumped periodically for new data to be captured. Somehow, the INCO had to find a way to get that data to the ground.

"We had to run the Shuttle's communication system in a different mode than we had planned," Castle said. This required a lot of replanning in a hurry. "We had only

The first flight of Spacelab, shown in the payload bay of Columbia during STS-9, produced more data than any previous mission. INCO had to get that data to the ground when the one and only relay satellite failed. NASA Photo KSC STS009-41071.

a few days to work out how to do this," Castle said. But by carefully scheduling the ground sites and recorders, they were able to manage.

By the end of STS-9, 720 billion bits (about 84 GB) of data had been transmitted by the communications system. This accomplishment led to Castle's selection to hang the plaque. As Castle said, "It all worked great, and the mission was a success." (Ref. 6)

Science Fiction to the Rescue

To send telemetry to the orbiting TDRS for relay to the ground, the Orbiter used a small (Ku-band) antenna dish. This dish was stowed in the payload bay for launch and entry. During those dynamic phases with the payload bay doors closed, communications were sent to TDRS or directly to the ground through four antennas located on the four corners of the cabin.

On STS-41G [October 1984], the Shuttle's Ku-band antenna malfunctioned. INCO Bob Castle, who later became a Flight Director, was on console, and responsible for this communication system.

"The antenna is a dish antenna much like a backyard satellite dish," Castle explained. "For the mission to be a success, the antenna had to point at a satellite in a higher orbit. It would not point correctly and was in fact, "banging around" erratically. It would bang hard side to side."

Then, not unlike people frustrated with home electronics, Castle's team came up with the ultimate solution. "I and my team figured out a way to stop the antenna from moving by unplugging a plug in the electronics." While this solution stopped the banging, the problem of maintaining good communications

Bob Castle, an INCO who later became a Flight Director, visits with his wife Sharon who served as a Payloads Officer during STS-61A. They were the second married couple to both work in the MOCR. (The author and her husband were the first.) NASA Photo 61A (S) 196.

with the Orbiter remained.

"I and the FAO [Flight Activities Officer's backroom] Pointers figured out how to point the antenna by pointing the Shuttle itself. This is like pointing a spotlight by steering a car instead of moving the spotlight." Castle credits this solution to an idea he came across in a science fiction novel. "I had read of this idea in an Arthur C. Clarke novel, *The Lost Worlds of 2001*, where he solved a very similar problem in the same way," Castle said. "By pointing the Shuttle instead of the antenna, we saved the mission." (Ref. 6)

During STS-41-G, the Ku-band antenna, shown here during STS-130, was stuck. INCO used an idea from science fiction to solve the problem. NASA Photo, 2010.

SHUTTLE INCO
STS-1 to STS-39

Name, first flight in MOCR

Edward I. Fendell, 1
Granvil A. (Gap) Pennington, 1
Alan Lee Briscoe, 1
J. E. Conner, 2
Robert E. Castle, Jr., 4
Harold Black, 6
Roberto P. Moolchan, 7
Richard W. Rodriquez, 7
Robert W. Harris, 41C
Raymond K. DePaolo, 41G
Al W. Baker, 51A
Edgar B. Walters, 51D
John F. Muratore, 51B

CHALLENGER
Joe W. Gibbs, 27
Charles J. Stafford, 33
Chris A. Counts, 41
Richard E. LaBrode, 37
Laura Hoppe, first flight between STS-77 and 107, (first woman)

(Ref. 48)

Gene Kranz became famous for his work as Flight Director during Apollos 11 and 13. He's shown here during Apollo 11 wearing his trademark white vest (made by his wife Marta) with fellow Apollo Flight Directors Pete Frank (behind) and Gerry Griffin. NASA Photo S69-26302.

FLIGHT DIRECTOR (FLIGHT)

Mission Control. That's the job of a Flight Director. All the other positions report to the Flight Director who considers the options and makes the decisions. These decisions could mean the difference between success and failure, life and death.

Flight directors oversee the work of about eighty strong-willed flight controllers. "Leadership is our job," Linda Ham, the first woman to serve as a Flight Director, said. (Ref. 14)

Like orchestra conductors, Flight Directors must not only have a thorough understanding of the systems "played" by each position but must also get the flight controllers to "play" as a team (with the astronauts) to achieve mission success.

Flight directors lead Mission Control and also represent the team to the public and to management, all the way up to the president. When Vice President Dan Quayle visited Mission Control in 1989, Flight Directors Tommy Holloway (left) and Bob Castle (right) showed him around. Michael Hawes (obscured by Holloway's arm), a former Payloads Officer who was with NASA HQ in 1989, also appears. NASA Photo S89-31322.

Most Famous Flight Director

Eugene (Gene) F. Kranz, who became the head of NASA Mission Operations (in charge of all flight controllers and Flight Directors), became famous as a Flight Director during Apollos 11 and 13. During Apollo 13, three astronauts were on their way to the Moon, and they had just done a standard stirring of the cryogenics tanks. (See the EGIL section.) Near the end of his shift, fellow Flight Director Glynn Lunney had just arrived to begin handover.

In those days, the vehicles did not have as many sensors as they do now. The first signs of trouble weren't all that clear. "For some reason, we had a high gain antenna go from narrow to wide gain," Kranz explained. "Basically, we trust the [onboard] systems." So, they thought the problem was something wrong with the data links or equipment in the Mission Control Center.

"The first minute we had the problem of which data we could and could not believe," Kranz said. "We spent about the first ten minutes troubleshooting instrumentation. By fifteen minutes, I recognized we had a significant mission anomaly. I asked Lunney to give center management a call and say we had a major mission problem."

About twenty minutes into the situation, the crew called down to say they were venting something and could see it out the window. Kranz said, "This is when my thinking went from 'this is an interesting problem' to one of survival. We had a major, major problem."

The astronauts had trouble maintaining attitude and were using high amounts of fuel. Things looked pretty grim. Kranz made a note in his log, "Are we in survival mode?"

They eventually discovered the venting came from an oxygen tank that had exploded. Oxygen in the tank had turned from liquid to gas and been ignited by a heater when the stirring occurred. The thrusters had come on trying to stop the vehicle from spinning and going off course like a balloon with a hole in it. The antenna had automatically gone to wide gain to keep from losing contact. But the real problem was that without oxygen to run the fuel cell power plants which provide electrical power to all systems, including life support, the crew were doomed.

The way Kranz dealt with the crisis has become standard procedure for all Flight Directors. His first rule of action: "Don't guess. We couldn't afford to make things worse." His second rule of action: "Secure those systems that we may need in the future." In this case that meant not using more battery power than necessary. And third: "Be conservative." In other words, hold onto resources to preserve as many options as possible.

Kranz called on his training experience for Apollo 9. For that mission, the training team had given them a number of complex rendezvous problems where they had to use the Lunar Excursion Module (LEM) as a lifeboat. The LEM was planned only to be used to fly to and from the lunar surface. The service module, which now had a giant hole in it, was supposed to bring them home. When it was determined that even with a massive power down, there wouldn't be enough power in the service module to last the time it would take to get back to Earth, Kranz decided the LEM lifeboat idea was their only chance to save the crew.

Kranz mobilized the team into action. "At this time of the mission, we had no LEM personnel on shift," Kranz said. "They had done their checkout and gone home. I requested them called back. I had them dust off the Apollo 9 lifeboat procedures.

"I initiated an emergency power down and stayed on top of the RCS [fuel] utilization because I was concerned we might not be able to sustain control of the vehicle due to the venting. About ten minutes later, thirty minutes into the problem, I called up all the network resources—all the remote sites." With the vehicle spinning around and unable to point the high gain antenna, these extra sites would help keep contact with the crew.

"I then asked the team to develop LEM power profiles to come home. I directed all the teams to work planning for coming back by a free return trajectory around the Moon rather than using the service [module] system. I no longer trusted it.

"Once we got to that status quo—secured systems, bringing people in, planning around the Moon, then we had shift handover. But I didn't sleep. I took the team, and we went down to room 210 [in Building 30] to establish what happened and determine implications for the remainder of the mission," Kranz said. "I worked four and half straight days with very minor breaks because I had the responsibility to produce the procedure package for the burn and entry. We had twenty hours to do the procedures, and eighty to get ready for entry."

The lifeboat plan worked, and the Apollo 13 crew made it home safely. For leading the Mission Control team that saved three astronaut lives, Gene Kranz received the Presidential Medal of Freedom. (Ref. 19)

Kranz continued as a Flight Director for Skylab, and then served as Mission Operations Director, and a living role model, for the teams of the first four years of Space Shuttle missions. He retired from NASA in 1994. (Ref. 23)

A Bad Command

The Shuttle Program offered its share of problems to be

Flight Directors L to R, Richard Jackson, Jeff Bantle (squatting), Robert Castle, Robert Kelso (squatting), and Gary Coen pose in front of the Mission Control historical marker. As of 1992, only thirty-two people had served as Flight Directors. NASA Photo S92-34411.

solved by Flight Directors. "On STS-32, I was the Flight Director when suddenly the Shuttle started spinning," recalled Bob Castle. "The crew was asleep and didn't wake up and notice. We lost communication with the Shuttle as it spun since the antennas weren't being pointed right. No one knew what had happened or why the Shuttle was spinning."

Like Kranz had done many years before, Castle collected pieces of the puzzle from his team. "The flight controllers in Mission Control reported to me what they saw," Castle said. "The DPS officer reported the computers were reporting an arithmetic error. This is an error a computer gets when it tries to divide by zero or something like that." What could have caused such an error?

"The GNC officer reported the Shuttle was firing jets as if it were trying to change attitude, but the attitude numbers were reported to, 'be crazy,'" Castle said. Having jets firing in all directions would waste precious fuel. Could they be reacting to a vent like on Apollo 13 or perhaps a jet that was stuck on?

The oxygen tanks looked fine, and there were no purges or dumps in progress that could account for the attitude change. Castle checked with the Propulsion Officer in charge of the jet engines. "The PROP

Flight Director Rob Kelso trained new Flight Director Linda Ham during a simulation. Ham became the first woman to serve as a Flight Director in 1992. NASA Photo S43-S-118.

Officer reported the jets were firing normally and being commanded to fire by the Shuttle's computers," Castle said. No jets were stuck on. It looked more and more like they were seeing the results of a computer problem.

The puzzle started to come together for Castle. He recalled, "We had just sent a navigation update, a command to tell the Shuttle's computers where it was in space." Castle called the FDO who had built the command. (FDO is in charge of the state vector updates during orbit operations when GPO is on call. GPO did the updates during the high-speed phases.) "The FDO could not confirm it was received correctly," Castle said.

"I put these pieces together and realized that a bad position command could cause all of these indications," Castle said. "The Shuttle was trying to point at the center of the Earth. If the navigation was bad, the Shuttle's computers would not know where the center of the Earth was (even though a person could look out the window and see it), and they were dividing by zero as they tried to figure it out."

Puzzle solved. What had happened was that the INCO had sent FDO's command just as the Orbiter was going out of communications with the ground. Some of the command had not been received.

Once the cause was identified, the problem could be corrected. "The CAPCOM managed to wake up the crew when the satellite picked them up again," Castle said. "The FDO then sent a new position command and this fixed the problem." (Ref. 6)

Solving problems like this was very satisfying to Castle. "Everyone should want to be a Flight Director," said Castle. "You are personally involved in virtually every aspect of a Shuttle flight. There's no place else (other than as an astronaut) where you have so much to do with the space program. You go from planning

On STS-32, Flight Director Bob Castle (shown here at his console) discovered the Orbiter was spinning out of control, the crew asleep and unaware of the problem. NASA Photo S32 (S) 252.

a flight to seeing it through, in person, and are part of its actual success. You don't plan and let it happen, you plan and DO it." (Ref. 6)

Castle served as Flight Director for 25 Space Shuttle flights and three International Space Station flights as well as serving as MOD for ten flights. He retired from NASA in July 2010.

SHUTTLE FLIGHT
STS-1 to STS-39

Name, first flight in MOCR

Neil B. Hutchinson, 1
Charles (Chuck) R. Lewis, 1
Don Puddy, 1
Tommy K. Holloway, 2
Harold M. Draughon, 2
Jay H. Greene, 3
John T. Cox, 3
Gary E. Coen, 3
Brock R. (Randy) Stone, 6
Lawrence S. Bourgeois, 7
Alan Lee Briscoe, 41D
T. Cleon Lacefield, 41G
Granvil A. Pennington, 41G
William Reeves, 51A
J. Milt Heflin, 51D
Charles (Chuck) R. Knarr, 51G
Chuck Shaw, 51J

CHALLENGER
Ron Dittemore, 29
Wayne Hale, 28
Robert (Bob) Castle, 34
R. (Rob) M. Kelso, 33
Phil Engelauf, 37
Linda Ham, 45 (first woman)

(Ref. 48)

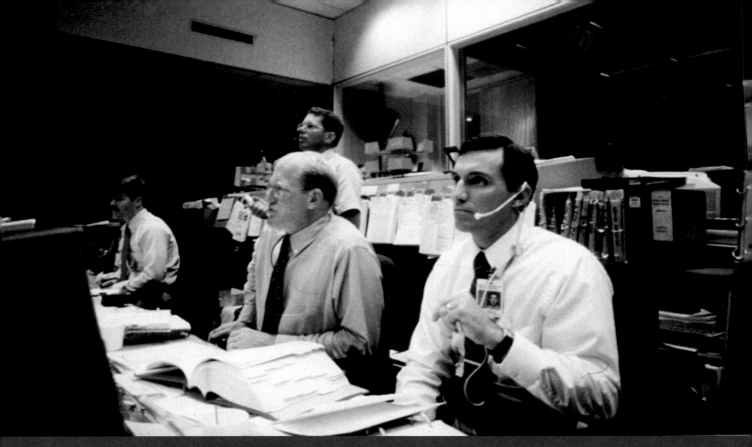

Astronaut Carl Walz (front) shared Capcom duties during STS-49 (1992) with fellow Astronaut Charles "Sam" Gemar. Astronaut Dave Leestma stands behind them and FAO Pete Hasbrook is seated at his console. NASA photo S49(s)-342-107.

SPACECRAFT COMMUNICATOR (CAPCOM)

To avoid confusion, all talking with the onboard astronauts came through one position. Although the Flight Director had the capability to talk with the crew, by tradition, an astronaut served as the spacecraft communicator during the Shuttle Program, relaying only those instructions approved by the Flight Director. VIPs such as Presidents also sometimes addressed the crew directly. The call sign, "Capcom," was a carryover from the days when spacecraft were called capsules.

Humor is a great outlet for the high-pressure job of Capcom. This "Capcom Tie" (worn by Astronaut Dan Brandenstein for STS-1) listed all the major calls made by the Capcom during a Space Shuttle ascent. Brandenstein donated the tie to the Smithsonian. NASA Photo S81-31688.

You Get the Rhythm

Astronaut Carl E. Walz was selected by NASA as a Mission Specialist in 1990. After his initial year of training, he became a Capcom. "We in the Astronaut Office, when we're not training to fly, support a number of different jobs," Walz said. Being a Capcom was one of those jobs. "From the standpoint of learning the systems, you can't beat working in Mission Control," he said.

Like other flight control positions, the Capcoms must be trained and certified. "You become certified to work as Capcom through doing a number of simulation flights," Walz said. "The simulations get you well versed in the malfunction procedures—that is, what you need to do to reconfigure the Space Shuttle if there's a problem. It gets you thinking, 'what can I do to make the Shuttle better if I had a problem.' Eventually you're certified and assigned to a shift for a given flight."

Astronauts who were pilots were normally assigned to the ascent, entry, and rendezvous shifts where they spoke pilot to pilot. Mission specialist astronauts were assigned to orbit shifts where most of the payload operations occurred. "During a mission, you learn the normal routine without all the failures thrown in for training," Walz said. "You get the rhythm for what normal operations are like."

Walz added, "Being a Capcom really gave me an appreciation for the folks that support our operations, for how hard they're working, and the tremendous amount of knowledge they have. They know the Space Shuttle to the Nth degree." (Ref. 37)

THE AVIATION ALPHABET

Acronyms speed communications with the crew. (NASA is a whole lot faster to say than National Aeronautics and Space Administration!) However, static often makes it difficult to distinguish one letter from another, such as "D" from "E." Pilots and airport flight controllers worldwide utilize a standard aviation alphabet to combat this problem. Using the alphabet, the acronym DPS, (Delta-Papa-Sierra) for Data Processing System, would not be confused with EPS (Echo-Papa-Sierra), the Electrical Power System. Ref. 2

A as in ALFA	J as in JULIET	S as in SIERRA
B as in BRAVO	K as in KILO	T as in TANGO
C as in CHARLIE	L as in LIMA	U as in UNIFORM
D as in DELTA	M as in MIKE	V as in VICTOR
E as in ECHO	N as in NOVEMBER	W as in WHISKEY
F as in FOXTROT	O as in OSCAR	X as in XRAY
G as in GOLF	P as in PAPA	Y as in YANKEE
H as in HOTEL	Q as in QUEBEC	Z as in ZULU
I as in INDIA	R as in ROMEO	

A Lot of Time Listening

Walz worked his first mission as Capcom in January 1992 during STS-42. "This was a Spacelab mission," Walz explained, "and the people at Marshall Spaceflight Center end up spending a lot of time talking to the payload crew."

The position in the Payload Operations Control Center at Marshall was called the Payload Communicator. It was usually filled by an alternate payload specialist familiar with the onboard experiments. "If there's a problem with the Orbiter, for example the reaction control system (which can be flight critical), then we would get involved and help troubleshoot," Walz said.

During Spacelab flights, the Capcom in Mission Control shares "talking" duty with payload communicators located in the Payload Operations Control Center in Huntsville, Alabama. Alternate Payload Specialist Rick Chappell served as a payload communicator during the ATLAS-1 Spacelab mission. NASA MSFC Photo 915415.

"Otherwise, we spend a lot of time listening," Walz explained. Listening to what? "Normally, there are two loops I'm intently listening to. The first is the air-to-ground loop, and the other is the Flight Director's loop. All the different front room operators talk to the Flight Director on the flight loop, so I listen to that and try to get a heads up on what crew actions are coming.

"When you call the crew, you want to include not only an action for the crew to perform, but you want to give them the reason why they're performing it. If we say, 'we want you to fire an RCS thruster,' we give them the whole story—'because the heaters are not operating and the thruster's getting cold.' Part of my job is to craft that story I'm getting from the flight loop."

Other calls the Capcoms made did not require input from the Flight Director or the rest of the Mission Control Team. "Acquisition of signal calls are always the same because we make them all the time," Walz said. "When we fly over the area near India, we don't have TDRS coverage, so about two minutes before, we tell the crew we'll be loss-of-signal for so many minutes, and we'll pick them back up at a specific time. These are pretty generic calls." (Ref. 37)

Simultaneous Conversations

Missions with a lot of interaction between Mission Control in Houston and the onboard crew are most exciting for Capcoms. STS-49 (1992) which included three EVAs to retrieve and repair the Intelsat satellite, was like this for Walz. "STS-49 was just loaded with activity," Walz recalled.

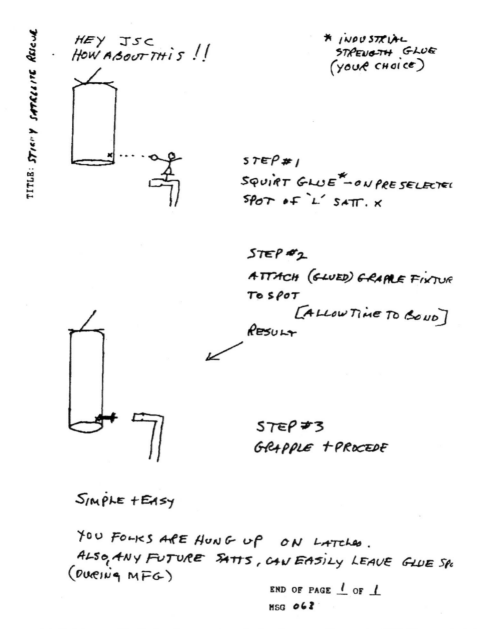

The Capcoms sent this "sticky satellite" plan for rescuing the Intelsat satellite to the STS-49 crew to brighten their day.

"After a second attempt failed to grapple the Intelsat, we stood down a day to figure out how we were going to approach the problem (to retrieve Intelsat). We were getting a lot of press [see PAO section], and people all across the country were faxing us different ways we might be able to capture the satellite. Some of these were good. We decided to fax one up to the STS-49 crew to brighten their day while they

To support the three-man EVA rescue of Intelsat on STS-49, three Capcoms were needed. L to R, are Capcoms Carl Walz, Sam Gemar, and John Casper. FAOs John D. Tolle and Pete Hasbrook (mustache) are in the background. NASA Photo S49 (S) 342.

were thinking, 'How are we going to do this?' We even got drawings from kids! I remember sending these up telling them that the whole country was behind them and was trying to brainstorm an answer to grappling the Intelsat. It was really fun."

On this flight, the Capcoms also got involved with the brainstorming. "I worked with Sam Gemar. As Capcoms, we had done a lot of training on the EVAs and were familiar with them. So if he was on console, I went to the meetings, or if he was going to a meeting, I would cover the console."

But when it was time for the EVAs, all the Capcoms were needed on console. "We had one rendezvous Capcom (John Casper] working a very pilot-intensive task. We were sending up information regarding how big a burn we needed, interpretation of onboard radar data,

The three-man rescue of Intelsat IV on STS-49 captured the imagination of the public. L to R, Astronauts Rick Hieb, Tom Akers, and Pierre Thuot capture the satellite for repair. NASA Photo S49-9120.

and flying-the-Shuttle kind of information," Walz said.

"Meanwhile, we also had these simultaneous conversations on airlock depressurization and spacewalk activities. Astronauts were reading down the parameters on their space suits because we were ready to turn the EVA crew loose (though they'd be flying in the payload bay with tethers). We were getting a lot of calls and people could channelize on the voices (EVA Capcom to the spacewalkers, Rendezvous Capcom to the commander and pilot)." (Ref. 37)

Spur-of-the-Moment Kind of Thing

The Mission Control Center was not open to visitors during flights. However, the news media, crew families, NASA employees, and VIPs were given access to the viewing room behind the FCR. Only on rare occasions were VIPs permitted into the Mission Control room itself. This happened when astronaut Carl E. Walz was serving as a Capcom.

"The President of the Ukraine came to visit during STS-49," Walz explained. "He was taking a tour of Johnson Space Center and Mission Control when he asked, through his interpreter, if he could talk to the crew."

Usually these "media events" were set up way in advance. But not this time. "It was totally unrehearsed and unexpected," Walz said. However, dealing with the unexpected is what Mission Control is all about. "So, we called the crew, who had gotten up just an hour before, and asked if it would be okay to talk to the President of the Ukraine." The onboard astronauts graciously agreed to the request.

"We put the President and his interpreter onto the air-to-ground loop," Walz said. "He was actually standing at the Capcom console. It was very interesting to meet a President of a large country." The President got what usually only Capcoms get to do during the mission, "a chance to talk with the crew."

He also got a chance to see the crew in real time. "We had TV downlink," Walz explained. "While the STS-49 Commander Dan Brandenstein was answering the questions of the president through the interpreter, you could see Dan talking on the TV."

Everyone was pleased with the results. "It worked out as if we had planned this all the

When the President of the Ukraine, Leonid Makarovich Kravchuk (left center), dropped in for an unexpected visit, Capcom Carl Walz let him talk to STS-49 Commander Dan Brandenstein from his console. Deputy Chief of the Flight Director's Office, Gary Coen, has his back to camera, head to the side. Because it was a surprise visit, no NASA photographer was present. This is a still from a video. NASA Photo S49 (S) 220.

Sally Ride became the first woman to be a Capcom during STS-2 in 1981. She is shown here with fellow Capcom Dale Moore. FAO Ben Ferguson is in the background. NASA photo s81-33963.

time," Walz said, "but it was just a spur-of-the-moment kind of thing." (Ref. 37)

Walz flew on three Shuttle flights, STS-51 in 1993, STS-65 in 1994, and STS-79 in 1996. He performed an EVA on STS-51. In 2001, he launched aboard STS-108 to the International Space Station where he served for six and a half months as part of Expedition 4, setting a U.S. spaceflight endurance record. He performed two spacewalks and returned to Earth on STS-111 in June 2002. After serving as the director for the Advanced Capabilities Division in the Exploration Systems Mission Directorate at NASA Headquarters, Walz left NASA in December 2008. (Ref. 50)

SHUTTLE CAPCOM
STS-1 to STS-39

Name, first flight in MOCR

Joseph P. Allen, 1
Daniel C. Brandenstein, 1
Henry W. Hartsfield, 1
James F. Buchli, 2
Frederick H. Hauck, 2
Terry J. Hart, 2
Sally K. Ride, 2 (first woman)
Steven R. Nagel, 2
S. David Griggs, 3
George D. (Pinky) Nelson, 3
Brewster H. Shaw, 3
Robert L. Stewart, 4
Michael L. Coats, 4
Roy D. Bridges, Jr., 4
Richard (Dick) O. Covey, 5
Bryan D. O'Conner, 5
Jon A. McBride, 5
Guy Gardner, 6
Mary L. Cleave, 6
John Blaha, 7
William Frederick Fisher, 8
Anna Lee Fisher, 9
Franklin Chang Diaz, 9
(first Hispanic)
Jerry Ross, 41C
R. N. Richards, 41D
Dave Hilmers, 41G
Ronald McNair, 41G (first Black)
R. C. Springer, 51A

R. M. Mullane, 51D
Dave Leestma, 51B
James Wetherbee, 51G
Fred Gregory, 61A
Anthony (Tony) W. England, 61A
C. Lacy Veach, 61A
L. Blaine Hammond, 61A
Shannon W. Lucid, 61A

CHALLENGER
John Creighton, 26
Kathryn C. Thornton, 26
G. David Low, 26
Mark Lee, 26
Kathryn D. Sullivan, 26
Pierre Thuot, 26
Marsha S. Ivins, 29
Michael A. Baker, 29
Ken Cameron, 29
Frank Culbertson, 29
Tamera (Tammy) E. Jernigan, 30
Steve Oswald, 30
Don McMonagle, 34
James Voss, 34
Story Musgrave, 31
Ken Bowersox, 31
Brian Duffy, 41,
Bob Cabana, 37

Note, this list may be incomplete.
No data was found for STS-51I, 51J,
61B, 27, 28, 33, 36, 38, and 39.
(Ref. 48)

During STS-3, Carolynn Conley became the first woman to be a Flight Activities Officer. NASA Photo S82-32942.

FLIGHT ACTIVITIES OFFICER

The Shuttle Flight Activities Officer was in charge of when the crew did what—in other words, the astronauts' schedule. Unlike most other positions, the FAO's "system" was therefore different for every flight. The FAO prepared or oversaw the preparation of all the flight documents. During the mission, the FAO rescheduled activities, figured out what direction the Orbiter needed to be pointed to do those activities, and, for the early flights that included a teleprinter, coordinated all printed messages sent to the crew.

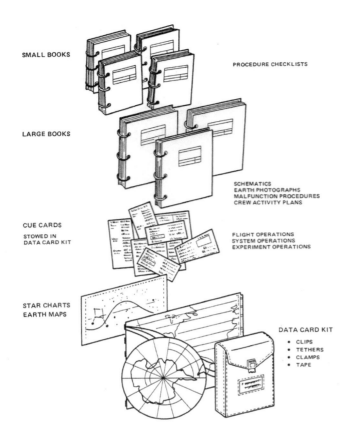

SMALL BOOKS

PROCEDURE CHECKLISTS

LARGE BOOKS

SCHEMATICS
EARTH PHOTOGRAPHS
MALFUNCTION PROCEDURES
CREW ACTIVITY PLANS

CUE CARDS
STOWED IN
DATA CARD KIT

FLIGHT OPERATIONS
SYSTEM OPERATIONS
EXPERIMENT OPERATIONS

STAR CHARTS
EARTH MAPS

DATA CARD KIT
• CLIPS
• TETHERS
• CLAMPS
• TAPE

The Flight Activities Officer was responsible for the complete set of cue cards, decals, maps and books which the crew took with them into space, called the Flight Data File. (Ref. 28)

The Best Laid Plans

Nothing ever goes just the way it is planned. Even though FAOs knew this, they had to prepare the best plan they could. They had to understand how everything fit together and be prepared for changes during flights.

On STS-3 in March 1982, Carolynn Conley became the first woman to hold the position of Flight Activities Officer. (During the first ten years of Shuttle operations, half of the FAOs were female, a higher percentage than any other Mission Control position.) Conley's planning team not only had to adjust the timeline, they had to rewrite the whole thing while the crew slept. "On this particular flight, the crew had been very busy and also adjusting to Space Adaptation Syndrome (SAS)," Conley said.

SAS is a general term for a broad range of symptoms that result from the body adjusting to the freefall environment. Two thirds of astronauts got "space sick" during their first few days of flight. (Ref. 3) Treatments and training helped astronauts overcome SAS, but Conley said, "During the early Shuttle flights, we still had no solution for SAS. We had to be very conservative with only two crewmen and a very demanding schedule of test objectives and experiments. We had to give the crew time to adjust to the space environment."

The plan for the third flight day was very crowded and complicated. Although the fourth day was also full of activities, these activities were considered less physically demanding. Therefore, the decision was made to swap the third and fourth flight day's plans.

"This [swap] might seem easy on Earth," Conley said, "but for a spaceflight this is a difficult task. The precession of the orbital path changed the relationship of the Orbiter to the stars, the targets on Earth, and the stars needed for test objectives." Like expecting to see Arizona out the window and seeing California instead, "it was not merely a matter of saying we'd take the activities from nine to noon on Day 4 and do them on Day 3. We had to adjust them all for where the Orbiter was with respect to Earth, the tracking stations, and the stars."

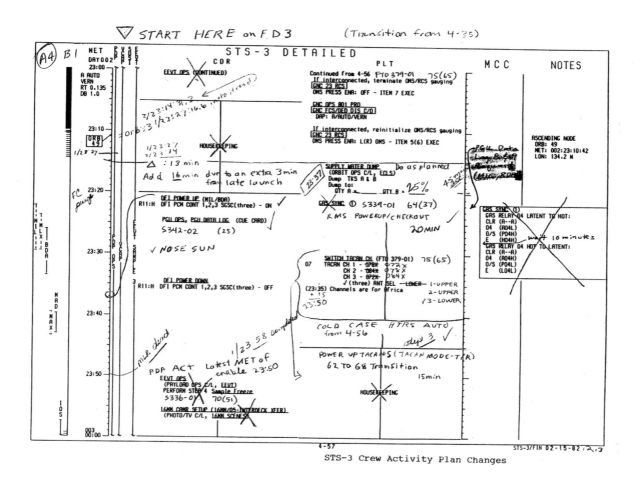

This page of the 1982 STS-3 Final Crew Activity Plan (Marianne Dyson's console copy) shows the numerous changes the FAO had to deal with when the flight team decided to swap activities from one day to another.

The timeline, which had been a year in preparation, had to be completely redone overnight. "We had to have the new plan uplinked to the crew before they woke up, so we only had about six hours to get the planning done!"

But Conley had help. "I was responsible for a team of about ten flight controllers in my backroom," Conley said. "I had people called PADS preparing the message, Pointers recalculating the maneuvers and attitudes needed for ground and celestial targets; and Timelines rescheduling the activities."

A computer planning system helped too. Although it hadn't been programmed to swap whole days, the STS-3 Timelines were able to create a new plan for the flight team to review. Each activity was electronically moved and checked against a new trajectory showing the day/night bands and ground station coverage for each orbit. Unfortunately, the crew weren't so lucky.

"We only had the equivalent of a typewriter printer for output in the Orbiter," Conley explained. "The crew would end up with several feet of instructions which they used to mark up their preflight books." The capability to send replacement pages to the crew wasn't available until a Text and Graphics System, like a fax machine, was put on board several years later.

Besides the backroom support, Conley received directions, corrections, and timeline changes from other console operators. "The FAO is the clearinghouse for all the inputs and configurations," Conley said. "During that six-hour period, we had to collect inputs from all the operators, put them into an integrated timeline, get it typed, distribute it to all the operators, and have the Flight Director finally approve it. Then the final message was formatted and uplinked." But even the uplink had to be carefully scheduled.

"The message couldn't be any longer than what the ground coverage would allow," Conley said. "We only had a six to ten-minute window of time because we were still using the old Apollo ground stations." [The first Tracking and Data Relay Satellite was launched into service on STS-6 in April 1983.] For STS-3, the message arrived just before the crew had breakfast.

"The crew didn't say it was good," Conley said, "maybe because they were more concerned at how long the message was, or how many instructions they had to do. But the real reward was when they didn't have difficulty following the instructions and didn't have questions. Then we knew we had communicated the replanning clearly for them." (Ref. 7)

Conley worked STS-4 (1982) and STS-8 (1983) as an FAO and then left NASA to work for a contractor. She was still working in the industry in 2021.

Fire in Mission Control

Usually, flight controllers deal with problems on the spacecraft. But during STS-5 in November 1982, Mission Control itself failed.

I (the author, Marianne Dyson) was an FAO when all the screens in Mission Control, and all the buttons on consoles suddenly went dark during STS-5. The following is what I wrote about this back in 1992:

"It was late at night in Houston, and the crew onboard the *Columbia* had already gone to bed. I was busy updating the timeline for the next day. I remember the Capcoms had just gotten ice cream and burritos when the Mission Operations Computer (MOC) went down.

"At first, I assumed the MOC had just overheated. But then GC reported

For the first few years of Shuttle operations, the crew (astronaut Richard Truly shown here during STS-2) had to hand mark document changes using instructions sent by the FAO to a line printer onboard. Eventually, replacement pages were sent using a Text and Graphics Systems similar to a fax machine. The photo is signed to the author (Marianne) who was Timeline in the FAO back room for STS-2. NASA Photo S81-3969.

a fire on the first floor! As the emergency lights came on, Flight Director Gary Coen said, "Secure the room. We are in lock down until further notice." This was not a case we had trained for!

"This flight was already different for me, because just a few days before launch, I found out I was pregnant. I thought of my unborn child and worried what might happen if someone decided we had to remain at our positions no matter what. We were on the third floor, and smoke from the fire was already making its way up through the vents—adding to the cigarette smoke in the room. (This was before smoking was prohibited in government buildings, and Coen was a chain smoker.) Building 30 had no windows and no exits except through the ground floor. I think we all felt somewhat giddy, not that we expected to die, but that the danger was real, and we were not in a simulation.

"Even though the power was out, the phone lines worked, so controllers could still talk to their back rooms and each other. INCO Bob Castle informed Flight that

During an STS-5, a fire in Mission Control took out communications with the Columbia in orbit. FAO Marianne Dyson (shown here during STS-4), who had just found out she was pregnant, was one of the controllers "locked" down inside the darkened, smoke-filled, third-floor MOCR. NASA Photo S82-33031.

all telemetry, command, and s-band voice capability were lost. The crew were asleep, but the lack of telemetry meant Mission Control could not monitor the Shuttle's systems or consumable status until the MOC was restored.

"GC Norm Talbot reported that it might take several hours to get the MOC back, depending on the severity of the damage caused by the fire. Security was preventing anyone from entering the building because of the smoke.

"Guidance Officer J.T. Chapman told Flight that unless communications with the Orbiter were restored by the next radio pass, an alarm would go off onboard to alert the crew that the Orbiter's state vector would soon 'go stale.'

Pearline Collector was the book manager of the Ascent Checklist, meaning she wrote the instructions in this book. She served as the Ascent Phase Specialist for STS-1, sitting next to FAO Bob Nute in the MOCR during the first hour of the flight. She is in the 1981 photo of women who worked in Building 4. Marianne Dyson photo.

FAO Carolynn Conley organized this group photo of all the women who worked in Building 4 in 1981. L to R, **Row 1***: Robin Cobbs, E.M. Eav, Rowena Burns, Maridene Lemmon, Iva Doyle, Monica Davis, Mindy J. Cohen, Cadie Howard;* **Row 2***: Anne Fox, Elaine Hilliard, Marianne Dyson, Anne Flippin, Panna Amin, Ann Johnson, Delores A. Couch, Teresa McDonald, Sandi Kovar, Anne Accola;* **Row 3***: Carolynn Conley, Helen Meester, Eva Hernandez, Maria Garza, Carmen Cruz, Pat Engel, Sharon C. Conover, Karen Flanagan, Georgie Huepers;* **Row 4***: Marie Gibson, Nina J. Weaver, Irma Cortez, Sara Beck, Pearline Collector, Nancy Harris, Patsy Smith, Susan Wilson, Mary Barth;* **Row 5***: M. Edson, M.E. Mabry, Ruth Ann Brinkworth, Barbara Hopkins, Dianne Murphy, Holly A. Barnes, Terry Stanford, Kathie Abotteen, Michele Brekke, Alice Van Gilder, Cheevon (Mi-Mi) Lau, Dianna Watson, Cynthia Nagy. NASA photo s81-28529C.*

"So besides the danger to us flight controllers, the crew were potentially in even more danger. The flight rule required an emergency deorbit while the state vector was still good enough to bring them down safely. Unfortunately, because of where they were in their orbit, they would be forced to land somewhere in Africa in the middle of the night.

"To avoid that nightmare scenario, INCO began work on a clever, but risky, plan to clear the onboard alarm: call the ground site on the phone and read them the hexadecimal code. Then the site operator would manually type it in and uplink it to the Orbiter.

"Meanwhile, me and my backroom team continued working on a teleprinter message listing changes to the next day's timeline that we would uplink to the crew in the morning. My backroom Pads operator, Jay Penn, called to say that he had confirmed that we could send that teleprinter message to the crew without going through the MOC!

"I happily shared this information with Flight. Though we couldn't use the teleprinter to update the state vector, we could at least send the crew a printed message and let them know what was going on.

Bill Holmberg (left) and Tucker Pierce (right) staffed the Timeline console in the FAO SSR during STS-1. Both later served as FAOs. Marianne Dyson photo, 1981.

"But we didn't want to wake them unnecessarily with the clatter of the teleprinter, so we waited, hoping it wouldn't be necessary. Fortunately, it wasn't. The MOC was rebooted after an hour and 15 minutes and came back online just minutes before the radio pass that, if missed, would have triggered the onboard alarm to wake the crew. Instead, Guidance sent the update, and the crew slept through the whole thing! I sure was glad to get home that night."

Because the fire happened during the night and NASA assured the news media that the crew were never in any danger, the event didn't get much press coverage. However, the fire may have been the most serious failure in Mission Control history.

SHUTTLE FAO
STS-1 to STS-39

Name, first flight in MOCR

Robert (Bob) H. Nute, 1
Elvin B. Pippert, 1
Ben E. Ferguson, 1
Henry J. (Tucker) Pierce, Jr., 2
Carolynn L. Conley, 3 (first woman)
Marianne J. Dyson, 4
William (Bill) R. Holmberg, 5
Cheevon (MiMi) Lau, 6
Charles (Chuck) R. Knarr, 7
Gerald L. Shinkle, 9
Robert L. Schaf, 41B
Phillip Lee Engelauf, 41C
Rick N. Jurmain, 51D
Tom L. Vollrath, 41C
Mark Rolwes, 41D
Karen F. Ehlers, 41G
Rick J. Hieb, 51A
Wayne E. Louis, 51A
D. Mark Maschoff, 51D
Neil A. Woodbury, 51B
Diane L. Freeman, 51B
Barbara A. Schwartz, 51I
Karen M. Alig (Engelauf), 51J
Anne F. Ellis, 61A

CHALLENGER
Gail A. Schneider, 27
Stephen S. Gibson, 29 (first Black)
Pete V. Hasbrook, 30
Michael Davis, 28
Nancy Jackson, 34
Debbie Jackson, 36
Tony Griffith, 31
Lee Wedgeworth, 38
Jeff Davis, 35
Fisher Reynolds, 37

(*Ref. 48*)

Technicians at Kennedy Space Center install a new auxiliary power unit for STS-31. MMACS Paul Dye made the call to scrub the launch when an APU began running wildly a few minutes before launch. NASA KSC Photo 90PC 605.

MECHANICAL, MAINTENANCE, ARM, and CREW SYSTEMS ENGINEER

MMACS, pronounced, "max"

In charge of everything from tires to eating utensils, Max was the all-around spacecraft mechanic and "handyman." If the payload bay doors stuck, the arm must be moved, the rudder was slow, a vent was clogged, or a camera needed repaired, Max was the one to call.

With such a long name, it's no wonder Mechanical, Maintenance, Arm, and Crew Systems Engineers preferred to be called simply, "Max." The MMACS position was created during the down time after the *Challenger* accident and was first staffed on STS-26. Max assumed responsibility for the mechanical and crew systems that EECOM and FAO had each, respectfully, overseen previously.

Paul Dye, shown here wearing a flight suit and parachute used for launch, spent over 2000 hours and two years training to become the first person to serve as a MMACS in Mission Control. The position was in charge of mechanical and crew systems. NASA Photo S89-42112-116.

MMACS Paul Dye, in charge of mechanical and crew equipment, got a turn in the hot seat of a Shuttle cockpit in 1988. He was in a different kind of hot seat as MMACS during STS-31 in 1990. NASA Photo S88-55216.

Wildly No Go

Even though the job was called "fun" by Paul Dye, the first MMACS Engineer, who became head of the MMACS section and then a Flight Director, was faced with some very serious decisions.

"During the first attempt to launch STS-31 (the Hubble Space Telescope)," Dye recalled, "one of the Shuttle's auxiliary power units, which provide hydraulic power for the flight control, began running wildly. It was less than five minutes prior to launch."

The auxiliary power units (APU) are not really auxiliary which means supplemental or reserve. All three APUs were needed for launch, one for each of the three main engines' valves and steering. Each APU was a turbine engine that converted the chemical energy of liquid hydrazine into mechanical shaft power. Without at least two of them, there was not enough "oomph" to steer the engines. (Ref. 28)

The crew was strapped in and ready to go, the television cameras rolling, and the flight control team

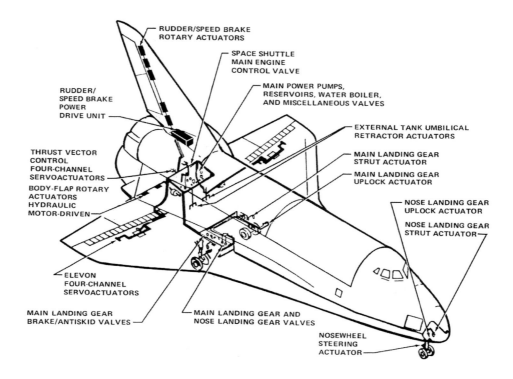

The Space Shuttle's hydraulic system was powered by the Auxiliary Power Units. Hydraulic systems are critical for the safe landing of the Orbiter. (Ref. 51)

anxiously awaiting the final countdown. The pressure was on Dye. "To be a great flight controller," Dye said, "you need to have an extremely fast mind—to rapidly understand changing situations and evaluate options in the wink of an eye."

A wink of an eye was about all the time Dye had that spring morning in 1990. And like a parent trying to decide, before the bus comes, if a slightly ill child will be okay or should see a doctor, Dye didn't have time to wait for symptoms to change. All he knew for sure was that "the signature [the readings of the sensors] was new and did not fit anything we had planned for."

In the final minutes before launch, the Flight Director asked each operator to give a go or no-go call before he decided to launch the Shuttle. Once the solid rocket boosters were ignited, it was too late to call it off. "Deciding to scrub a launch in the last couple of minutes is probably one of the toughest calls to make," Dye said. "It is a difficult judgement call. It has taken so much to get to that point that you really want to go if at all possible."

Turning the vehicle and support teams around for another try also cost millions of dollars. However, considering this was the first launch attempt for the Hubble Space Telescope, a payload that would have been ruined by a launch abort, no one wanted to risk launching without being completely certain it was safe. "We were not sure what was going on," Dye explained. There wasn't enough time to do a complete analysis of the problem, only time to decide if the APU was safe or not safe for launch.

"Through quick evaluation of strip chart data, and discussion of options, we (me and my back room) decided the risk of launching was too great," Dye said. "It was a gut-level call on my part, and I was responsible for calling the system, 'No-Go,' and scrubbing the launch."

But once the countdown clock had stopped, the cameras were turned off, and the astronauts had climbed out to try again another day, the workers at Kennedy Space Center removed the questionable APU. "When they took the unit apart," Dye said, "they found a broken valve seat." (Ref. 9)

This failure was absolutely serious enough to justify Dye's call to scrub the launch. But even if the APU had been found to have a minor problem, when people's lives are at stake, it was always better to be safe than sorry.

Nothing Compares to Console Work

Back in 1992, Dye said that "being part of the STS-26 Ascent/Entry team was probably the highlight of my career to that point. We trained so long together," Dye recalled, "and well, the last time anyone tried to launch one, it blew up. Very rewarding to hear MECO (Main Engine Cutoff) on that one."

He added, "You can't get me to give up hands-on work. Period. Even when I was the (MMACS) section head, I maintained my console proficiency and covered the console when no one else was available. You never get bored—it's different every day. It's the closest thing to playing a game for a living as you can get. In other words, I haven't found anything yet that would make me leave console work. Maybe if NASA gave me my own T-38 [training jet] to fly around for a living, I could be convinced to give up console altogether, but otherwise . . . ." (Ref. 9)

NASA didn't give him his own T-38, but Dye went on to become the longest-serving Flight Director, working as Flight in Mission Control from 1994 through his retirement in 2013. He is also a commercial pilot and remains actively involved in building, flying, and writing about experimental aircraft.

MMACS Paul Dye shown here on console in Mission Control in this undated photo. He later became NASA's longest-serving Flight Director. NASA Photo courtesy Paul Dye.

SHUTTLE MMACS
STS-26 to STS-39

Name, first flight in MOCR

CHALLENGER
Paul F. Dye, 26
David F. Thelen, 26
Richard C. Poch, 26
Kevin R. McCluney, 27
Robert C. Doremus, 29
James W. Medford, 28
William C. Anderson, 28
Ladessa C. Hicks, 36
(first woman, first Black)
Alan Bachik, 41

(Ref. 48)

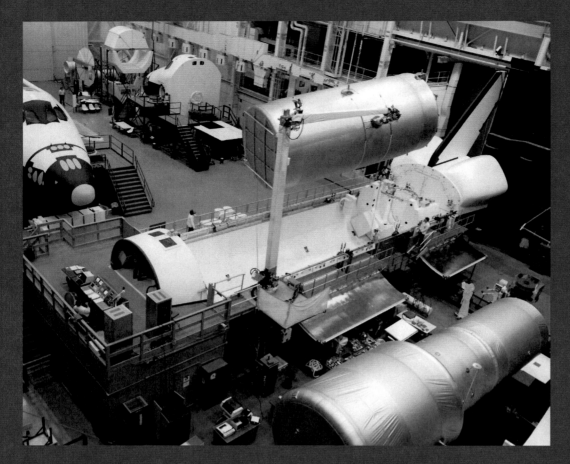

Astronauts and flight controllers tested and practiced arm procedures using a full-size Shuttle mockup in Building 9 at Johnson Space Center shown here in 1984. Designed for space, the arm couldn't lift its own weight on Earth. Therefore, giant helium balloons (shown) and Styrofoam and balsa wood mockups were used for training. NASA Photo S84-30844.

PAYLOAD DEPLOYMENT and RETIEVAL SYSTEM (PDRS) SPECIALIST
Remote Manipulator System RMS

The Payload Deployment and Retrieval System Specialist was in charge of what was probably the world's most complicated joystick. This joystick was formally called the Remote Manipulator System (RMS), but most people called it the "arm."

The robotic arm was built by Canada and first flew on STS-2 in 1981. It was used to grab payloads the size of school buses and remove them gently from the payload bay (i.e., deploy them); or to nab payloads already in space (i.e., retrieve them) and put them in the payload bay for return or repair. The arm also was often outfitted with a small platform for astronauts to stand on securely while they worked, like electricians use cherry pickers to work on power lines.

The arm was equipped with multiple cameras that were used to inspect payloads and the Orbiter itself. But the arm was not long enough to reach and inspect some parts of the Orbiter.

[Author note: the following paragraph was added in 2021 to replace outdated information on a planned payload deploy system that never flew.]

After the *Columbia* accident in 2003, which was caused by debris during launch damaging tiles on the wings, the Orbiter was outfitted with the Orbital Boom Sensor System (OBSS). The OBSS, first used on *Discovery* in 2006, was a sort of selfie stick. The OBSS was mounted on the starboard side, and the RMS was on the port side. The RMS would pick it up and use it to reach around to the belly of the Orbiter to check for damage to the tiles. (Ref. 52)

Who Ya Gonna Call?

While doing a routine survey, the crew of STS-41D (August 1984) discovered a problem. PDRS Officer Ronald J. Zaguli, then in a backroom position, witnessed the event.

"We were having trouble dumping [water], and the temperatures of the [wastewater] nozzle were decaying rapidly indicating there was a potential ice growth," Zaguli said. "So, we flung the arm over the side to take a look at the nozzle, and lo and behold we saw a rather large chunk of ice grown out from the side of the vehicle. It started getting real cold at the nozzle and the flow rates dropped to almost nothing because it [the nozzle] was occluded by a chunk of ice."

The Payload Deployment and Retrieval System (PDRS) Specialist was in charge of the robotic arm, shown here during its first flight on STS-2 in November 1981. The fifty-foot mechanical arm was built by the Canadians. NASA Photo STS002-13-226.

Wastewater from washing and urinating were stored in a tank and dumped overboard once a day. If wastewater couldn't be dumped, the crew was stuck going to the bathroom in sealed bags. While that would be inconvenient for the STS-41D crew, the ice posed a more serious danger to the Orbiter itself.

"There was a concern because on STS-11 [41-B, February 1984], we found ice and had not broken it off," Zaguli said. "During reentry it [heated,] broke off and smashed into the OMS pod, causing damage and schedule impacts." This was an experience the flight team determined to avoid repeating if at all possible.

"There were a number of different options discussed," Zaguli recalled. "The RMS team went off to work if we could use the arm to physically push that ice chunk away from the vehicle." The fifty-foot mechanical arm had never been used for something like that before.

"The Mission Management Team, including representatives from engineering, the Cape, the payload customers, and the Mission Operations Directorate, got together," Zaguli said. "They heard all sides, weighed the potential of arm failures against the more important goal of breaking off the ice."

The arm couldn't even lift its own weight on Earth. Its "fist" consisted of a rather delicate round "end effector." A circular ring "lined" the inside of the cylinder and rotated to cause three wires to close around a "grapple fixture," sticking out from a payload like a doorknob on a pole. The grapple fixture was "seen" via a TV camera on the end effector. (Ref. 27)

"The RMS team went off and started developing procedures to apply force and hopefully push the ice away," Zaguli said. There was no time to waste (no pun intended!).

"One of the unique things done by the superb simulator team folks was to use a cone to simulate the ice," Zaguli said. But the simulator and PDRS teams were not the only ones involved in testing the new use for the arm. "Astronaut Sally Ride tried the procedure in the Shuttle Engineering Simulator in building 16."

Like trying to tie a sash behind your back for the first time, the new skill encountered a few snags. "Because we were placing the arm so close to the vehicle, which was not something normally done," Zaguli said, "we actually closed the payload bay door part way to give us more clearance." (The arm had to be in its cradle for the payload bay doors to close. If for some reason the arm couldn't be stowed, it could have been "amputated" via an explosive charge at the shoulder joint. (Ref. 27)

While the PDRS, simulator, and astronaut teams were experts on the arm, this aspect of the procedure required engineering detail not available in the simulator. "We woke up some folks at the Cape to take measurements of the clearance between the wing and nozzle to make sure we wouldn't do any damage by bringing the arm in there," Zaguli said. "We didn't want to pound anything but the ice with that 'fist!'"

Just like a human arm, the RMS had joints at the shoulder, elbow, and wrist. These joints were commanded to move separately or in combinations. The sequence needed to hit the ice had to be carefully figured out. (Ref. 27)

Meanwhile, the rest of the flight control team was also busy "The team tried a number of different things [before using the RMS]," Zaguli recalled. "We shook the

The end effector, an engineering test unit shown here, on the Shuttle's robotic arm contained three wires that wrapped around a grapple fixture (center pole) when the inner ring rotated. Marianne Dyson photo, 1996.

On STS-41D, a camera on the arm took this picture of ice blocking the Orbiter wastewater vent. It was feared that unless the ice were removed, it would break off during entry and damage the engine pods. NASA Photo S14-33071, 9-2-84.

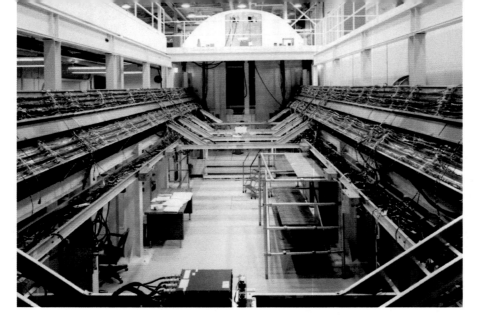

PDRS Specialists practiced using the software in the Shuttle Avionics Integration Lab (SAIL) which was located in Building 16 at JSC. "The simulator had a graphical mockup of the robot arm," PDRS Specialist Ron Zaguli said. NASA Photo S79-3714.

Orbiter by firing the jets. We tried to bake the ice off [by pointing toward the sun]." But the ice remained stubbornly attached.

The flight team then turned to the PDRS team for a better solution. That's when something unusual happened to Zaguli. "Even though I was a backroom operator," he said, "I was asked to come to the front room and tell the Flight Director what was going on. I had spent most of my shift in the simulator working with a crew to develop procedures. I had been working about twelve hours, and Mr. Kranz (MOD) said something to me about being in the heat of battle." Zaguli reported the results of the simulator tests to the Flight Director.

But being conservative and aware of the potential risk to the arm, "They decided to do more of the 'shake and bake'[first]," Zaguli said. "We had done all the work to be ready [to use the RMS option], so since I was exhausted, the front room sent me home." But that wasn't to be the end of it for Zaguli.

"When I got home, I couldn't sleep," Zaguli admitted. "So I turned on NASA [TV] Select. It turned out the shake and bake hadn't worked. I saw

PDRS Specialist Mark Ferring (left) discusses the STS-41D ice problem with FAO Chuck Knarr (in uniform), astronaut George (Pinky) Nelson (blonde), and NASA management. They decided to use the arm to knock it off. NASA Photo S14-3208.

the arm down in position, and I heard Sally [Ride] say the decision had been made to do the [RMS] ice removal. Then I didn't sleep at all. I was glued to the TV!"

The procedure Zaguli and the others conceived and tested worked, removing the ice, estimated to be about twenty inches long. The Orbiter was secured from an impact during entry, and the arm wasn't damaged.

Zaguli said, "From the inception of the procedure to doing it was only about twenty-four hours. It was an outstanding effort by everyone working together." Although Zaguli, like most flight controllers, was quick to praise and give credit to the team, his outstanding work was not overlooked by the Flight Directors and management.

"The next day, entry day, the Flight Director asked me and Mark Ferring [a fellow PDRS team member] to hang the mission plaque," Zaguli said. "We were the first backroom operators to do this." Thereafter, he and his PDRS team were called the "Ice Busters." (Ref. 38)

Zaguli was promoted to the front room position of RMU for STS-61B in 1985, and then worked as PDRS for STS-27, 32, and 37. He transitioned to the Space Station Program Vehicle Office in the mid-90s. He was awarded a patent for an "Attachment Device" by the U.S. Patent Office in 1994.

SHUTTLE RMU/ RMS/PDRS
STS-2 to STS-39

Name, first flight in MOCR

William D. Reeves, 2 (RMS)
Dale E. Moore, 2 (RMS)
Robert (Bob) R. Kain, 3 PDRS Phase Specialist
Jack Knight, 4 (RMU)
John D. Blalock, 4 (RMU)
Albert Y. Ong, 4 (RMU)
Arthur L. Schmitt, 4 (RMS)
Rodney L. Lofton, 5
Robert E. Anders, 6 (RMS)
Kitty A. Havens, 6 (RMU) (first woman)
Pramod Kumar, 8
Mark J. Ferring, 8
Karen E. Ehlers, 8
George H. Ulrich, 9
Edward J. Ripma, 41B (RMS)
Dave P. Huntsman, 41G
Gary Pollock, 51C (RMS)
Arthur Schmitt, 51C (RMU)
J. Steve Mclendon, 51B
Keith A. Reiley, 61A (RMU)
Ron J. Zaguli, 61B (RMU)

CHALLENGER
Ronald L. Farris, 27 (PDRS)
David S. Moyer, 41
Don Pallesen, 41

Note, this list may be incomplete. The position was not assigned for all flights. (Ref. 48)

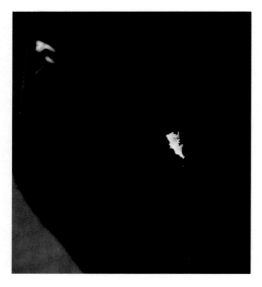

This large chunk of ice was knocked free of the Orbiter by the Shuttle's robotic arm on September 2, 1984. NASA photo S14-33071.

The 1984 movie "Ghostbusters" starring Bill Murray, Dan Aykroyd and Sigourney Weaver, inspired the name and logo of the PDRS Team that "busted" the ice off the Space Shuttle Orbiter that year on STS-41D. (Ref. 49)

The solid rocket boosters were impressively large and powerful. Lying down, three full-sized buses could be parked end-to-end inside one. Standing up, a booster was about the same height as the Statue of Liberty though it weighed about three times as much. Each case was designed to be reused about twenty times. (Ref. 33) Photo courtesy Thiokol Corp.

BOOSTER ENGINEER (BOOSTER)

The Shuttle Booster Engineer oversaw the operation of the three Shuttle main engines and the two solid rocket boosters used for launch. A lot happened in the first eight minutes of ascent as the combined thrust of the main engines and the boosters pushed the Orbiter into space. In that short time, the Shuttle went up about seventy miles and downrange (horizontal) over eight hundred miles—halfway across the Atlantic Ocean. (Ref. 28)

The reason Shuttle boosters were called solid rockets was because they used a solid rather than a liquid fuel. Five tons of propellant burned in one second—enough to make a "block" four feet (1.2 m) square and five and a half feet (1.7 m) tall. Cured propellant looked and felt like a hard rubber eraser because the curing agent was a synthetic rubber. This fuel was mixed in six-hundred-gallon bowls (shown here) at the factory in Utah. (Ref. 33) NASA MSFC Photo 885388.

This picture was taken at the rollout of the first Shuttle External Tank (ET). One tank held one and half million pounds of liquid hydrogen and oxygen which was used by the Orbiter main engines in about eight minutes. (Ref. 28) The ET was not reused. It burned like a meteor falling back to Earth. For the first two flights, the ET was painted white, but subsequent flights left it its natural orange to save money and weight. (Ref. 23) NASA MSFC Photo 783349.

Booster Franklin S. Markle, III, with a degree in electrical engineering from Rice University, went to work for the space program after friends told him how exciting it was. He said, "The rewarding part of this job is making sure everybody's flying with safe engines and safe solid rockets. We know that's critical after 51L [the *Challenger* disaster]. A lot of work goes into getting engines ready. We support a lot of reviews to make sure the Shuttle is going to fly safely."

One of these reviews took Markle to see a solid rocket test at Thiokol's plant in Utah. "The raw power of these devices is just incredible. On the test stand, the nozzle exit stays at the same point, as opposed to moving away like it does for a launch, and it's like looking into an intensely bright and hot sparkler with all those metal particles in the plume. When you feel the ground shaking, you know God has a hand in every launch that's successful."

What did a Shuttle Booster Engineer do? "I am kind of the choreographer of launch," Markle explained. "I let everybody know things are happening on time that are supposed to happen."

Markle said he scanned multiple display readings and warning lights during the high-speed ascent phase. "If the computer has a problem, the crew gets a fault message onboard, we get a fault message on the bottom of our displays, and we also have multiple indications depending on the scenario," he said.

"About thirty seconds into ascent—what we call the thrust bucket—where the engines have to throttle back to keep the atmosphere from over pressurizing the vehicle, I verify the thrust steps. First comes 'throttle down,' then after the bucket, 'throttle up to 104%.'" Markle said the engines must throttle down, "to keep the Orbiter from coming apart.

"My next call [to the Flight Director who then calls the crew via the Capcom]," Markle said, "is 'three-g [three times the force of Earth's gravity] throttling.' This is when the engines must throttle back or structural damage can occur to the vehicle by overaccelerating it. I don't know if the wings would crack, or the tank collapse," Markle noted, "but it's not designed for more than three and a half g's, and we only take it to three.

"The next call is 'MECO,' main engine cut-off, followed by 'MECO confirmed.' This means the Orbiter general purpose computers verify all three main engines have been shut down. It is very critical that the Orbiter computer tell the main engine computer what throttle settings to go to and to tell it to quit [running]. If an engine runs out of gas while it is still running, the Shuttle will blow up," he said.

"Then I verify the valves have cycled properly on the Orbiter side to blank off the External Tank,

The Space Shuttle was constrained to launch East so the solid rocket boosters (shown here being recovered) fell into the Atlantic Ocean and not on populated areas. The boosters, which usually splashed down about one hundred and forty miles (225 km) off the coast of Florida, were cleaned and reused. (Ref. 33) Photo courtesy Thiokol Corp.

Booster Engineer Frank Markle observed a solid rocket motor test like the one shown here. The plume of the flame is up to five hundred feet (152 meters) long. The burning fuel produces huge quantities of hot gas which travel at about five times the speed of sound or three times the speed of a high-powered rifle bullet. Photo courtesy Thiokol Corp.

isolating the tank [which re-enters the atmosphere and burns up] from the Orbiter. Once this is done, I make the 'ET sep' [separation] call and later monitor the dumping [overboard] of the residual oxygen."

Some people may wonder why this extra fuel is dumped into space instead of saved for later use. Markle explained, "If we didn't get rid of the oxygen, being a cryogenic propellant which is very, very cold, it would boil and expand, creating too much pressure in the oxygen pipes. It's the same with the excess hydrogen. [But,] the hydrogen is [even more] dangerous because when it gets back in the atmosphere, in contact with oxygen, it burns."

Though Markle was not on duty during the *Challenger* explosion, he said the Booster Jerry Borrer would not have had any warning of what was about to happen. "Indications were that there was a slight difference in the two chamber pressures on the left and right boosters, but that isn't abnormal. There was no indication something was wrong until the Shuttle blew up and the data stopped," Markle said. Unlike the main engines, there is no way to shut down a solid booster once it is burning. "It was

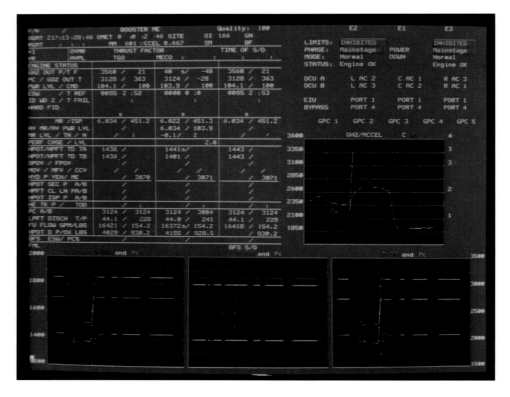

During launch, Boosters monitored the condition of the Shuttle's main engines via console displays like the one shown here from 1988. The plots at the bottom showed the temperature and chamber pressure for each of the three main engines. NASA Photo S88-45277.

catastrophic, nearly instantaneous."

The *Challenger* accident was blamed on a faulty design of the solid rocket booster which was then redesigned and extensively tested. The Shuttle Program never had another solid rocket booster fail.

Markle said that for the ascents he's worked, the engines have performed thankfully within normal operating ranges. "It's a miracle how a machine as complicated as the Space Shuttle engines, especially three of them, can do what they're told and do it when they're told flight after flight. The major feeling you have sitting there on console is the incredible power of the machine." (Ref. 21)

Could Have Gone to Africa

However, even a nominal flight could test a flight controller's nerves. "On STS-44 [November 1991], we had a problem that got everyone's attention," Booster Engineer Markle recalled. "The way we determine the amount of thrust on an engine is to look at the chamber pressure. The main engine computer needs to know if it's getting the right thrust and is burning propellant in the right ratio [mixture]. It uses the pressures from four transducers placed in the combustion chamber."

Booster noted the problem during the first few minutes of the flight. "One pair biased due to contamination," he said. "There was a blockage of a little tube that goes from the chamber to the transducer. The transducer biased high. The engine [computer] uses an average of the four transducers, and it falsely thought the thrust was something it wasn't, that it was higher than it should be. To get the average back where it was supposed to be, it [the computer] had to throttle the engine down."

The engine therefore wasn't producing as much thrust as it normally would have. This was a definite cause for concern. "Engine performance is very critical," Markle said. "Certain performance cases can put you in the water."

However, from numerous training sessions, Markle didn't think this case would result in a ditch. "But," he added, "it got everybody's eyebrows up, because if it had gotten bad enough, it could have caused us

Photo left: *During the STS-44 launch, Booster Frank Markle saw a main engine, like the one being tested here, throttle down due to a blocked tube. The engine nozzle is ten feet (3 m) long and eight feet (2.4 m) in diameter, about the size of a small office. These nozzles must take the full blast of hot gases time and again, so they are honeycombed with over one thousand tubes (with hydrogen flowing through them) for cooling.* Photo courtesy Rocketdyne. (Ref. 32)
Photo right: *The Shuttle Booster was responsible for the solid rockets on either side of the orange external tank and the Orbiter's three main engines—seen here firing during the launch of STS-40 in 1991.* NASA photo STS-40-9133610.

Jenny Howard (later Stein), shown here in a support room position during STS-1, was the first woman to serve as Booster. Others in the photo, front to back are: Unidentified man, Cleon Lacefield, Keith B. Lawler, unidentified Air Force officer (standing), Ray Halyard, Bill Murray, Jerry Borrer (standing). NASA photo s81-29621.

to go to Africa to what's called an intact abort site."

But fortunately, it didn't get that bad. "The blockage cleared around four minutes after launch," Markle said. The mission continued to a nominal orbit and a successful flight. (Ref. 21)

Markle served as Booster through STS-55 in 1993. He was promoted to lead the Communications System Group in the mid-90s and then moved into the EVA, Robotics, and Crew Systems Operations Division and became chief of the Robotics Operations Branch of the Neutral Buoyancy Lab Office (according to the JSC 2006 phone book).

SHUTTLE BOOSTER
STS-1 to STS-39

Name, first flight in MOCR

Jack A. Kamman, 1
T. Cleon Lacefield, 2
Jerry L. Borrer, 2 (on console for Challenger 51L)
Baldwin G. Fitzgerald, 4 (Entry)
Jenny Howard (Stein), 5 (first woman)

CHALLENGER
Thomas W. Kwiatkowski, 27
Mark R. Jenkins, 32, 33
James M. Dingler, 36 (Entry)
Frank S. Markle, 39 (Entry)

(Ref. 48)

During Apollo-Soyuz in 1975, Pete Frank (left) and Gene Kranz (right) served as Mission Operations Directors. NASA Photo S75-26593.

MISSION OPERATIONS DIRECTOR (MOD)

The Shuttle Flight, later Mission, Operations Director (FOD/MOD) was a Flight Director with the big picture. MOD knew how a particular Shuttle flight fit into the overall program and policies of the agency. "The MOD's first job is to keep the outside world off the back of the Flight Control Team," said Eugene (Gene) F. Kranz, MOD for the first four years of Space Shuttle flights.

"The MOD's second job is to act as the agent of the Flight Control Team in keeping the program management, NASA Centers, technical directors, and public affairs appraised on the status of the mission and provide the rationale for the direction the team intends to take.

"The third job of MOD is to listen to the outside world and see if they have something useful to bring to the attention of the Flight Control Team. For example, in the early Shuttle Program, the data I got from outside was so compelling that I had to instruct the Flight Director on two occasions to change the direction he was going." (Ref. 19)

FOD Gene Kranz surveyed the Flight Control Room from his position in the back row just after the landing of STS-2. Also pictured, Pete Frank (left, hand on console), Flight Directors L to R, Neil Hutchinson (beard), Harold Draughon, and Don Puddy. NASA Photo S81-39506.

They Wanted to Talk About It

"Leadership—the ability to focus the imagination, the energies, the talents of a group by force of example, talents or personal qualities to evoke images and expectations—to develop choices that give images substance and finally to achieve the established objectives."

This quote was prominently displayed above the grease board used for mission planning in Kranz's office in Building 1 of the Johnson Space Center. More than any other person, the former Apollo 13 Flight Director (see Flight Director Section) shaped the conduct of Mission Operations through his leadership as the head of the Mission Operations Directorate. He served as MOD for Skylab and the first seventeen Shuttle flights.

Kranz recalled an experience where a manager was needed as a broker between the flight control team and upper management. "I was MOD during STS." During the first day of the flight in November 1982, "we had indications of high PH readings on fuel cell water. We were concerned this might indicate a possible membrane breakthrough. The effect was the potentially

MOD Gene Kranz (left) with Flight Director Don Puddy beside him explained Mission Control's decisions to the media at a press conference during STS-2 in 1981. NASA Photo S81-31676.

Leading by example, MOD Gene Kranz sports his motto for Mission Controllers, "Tough and Competent," in this photo from 1987. NASA Photo S87-44346.

catastrophic mixing of hydrogen and oxygen in the fuel cell." The crew's lives were at stake.

"We'd already determined the proper course of action was to shut down that fuel cell," Kranz said. "This left us with the next action, a powered-down entry. We wanted to come home as soon as practical because we did not know the cause of the failure. We did know we had extremely limited experience and pedigree in our power down procedures." STS-2 was the second flight of the Orbiter *Columbia*, and the first Shuttle flight for the crew (Joe Engle and Richard Truly). Many of the contingency procedures had only been tried in simulations.

"We got into this long telecon with [NASA] Headquarters," Kranz said. "We had just gotten a new NASA Administrator. Kraft [Johnson Center Director] was there and told Headquarters what our intention was for early mission termination."

Because of budget constraints, the flight test program had been cut to only four flights. Knowing a replacement flight for STS-2 would cost around a billion dollars, it was understandable that NASA approach the early termination with care. But the longer the flight continued with only two fuel cells, the more risk there was that another would fail and put the crew and spacecraft at risk.

"Headquarters indicated they wanted to talk about it," Kranz said. This frustrated the former Apollo 13 Flight Director who wanted a quick decision. "This was early in the afternoon. We talked for about three hours, and it was time to put the crew to bed. I wanted to tell the crew we were going to enter, and we would have a partially powered-down entry. Our objective was to be in a posture to sustain another fuel cell failure." There were three fuel cells, at least one of which was required to land the Orbiter.

"I agreed with the Flight Control Team's conservative position, relative to the power down procedures," Kranz said. "I knew the potential implications of another fuel cell failure. I knew how they thought, and I carried that message to the briefing with the Headquarters team.

"While people on console worked the problem, I kept Headquarters appraised of when we had a deadline for decisions. I took the pressure off the Flight Director so he could work the fuel cell problem and do what he had to do while I did the politics," Kranz said.

Eventually the deadline for a decision arrived. "It was time to get the team planning entry and working through the checklist that night," Kranz said. "But we still couldn't make any headway with Headquarters. So, I told Kraft I recognized we had a problem convincing Headquarters this was all necessary, but we couldn't afford to delay any further on this decision. I told him I expected him to sell Headquarters [on the plan] because I was going to tell the team exactly what we were going to do. Eventually, that's what was done."

The crew were informed they would return to Earth after one day in orbit. In that one day, the crew

Flight Director Tommy Holloway, shown here during STS-3 in 1982, was the first person to serve as MOD after Gene Kranz. NASA photo s82-28714.

managed to perform almost all of the major flight objectives. The remaining two fuel cells did not fail, but the weather at Edwards Air Force Base forced the first, and only, Shuttle landing at White Sands, New Mexico.

Kranz, recipient of the 1970 Presidential Medal of Freedom, 1986 Space Flight Award, the 1988 Robert Gilruth Award, and the 2007 Rotary National Award for Space Achievement, talked about his real reward. "Just being a service to the Flight Control Team. If you've done a good job as MOD, you've allowed the team to get their work done without too much disruption. But the real reward is just the opportunity to be present in a room where you've got a magnificent team doing the marvelous things they do. It's like watching over the shoulders of a Michelangelo or a Da Vinci, being present when history is written." (Ref. 19)

Kranz retired from NASA in 1994. His memoir, *Failure is Not an Option,* was published in 2000. It was a New York Times bestseller and was adapted for the History Channel in 2004. He was portrayed in various films including *Houston, We've Got a Problem* (1974), *Apollo 13* (1995), HBO miniseries *From the Earth to the Moon* (1998), an episode of NBC series *Timeless* (2016), and in the series *For all Mankind* (2019). A character in the videogame, *Kerbal Space Program,* and the Eugene Kranz Junior High School in Dickinson, Texas are named after him.

SHUTTLE MOD
STS-1 to STS-39

Name, first flight in MOCR

Eugene F. Kranz, 1
Tommy W. Holloway, 51G
Don R. Puddy, 61A

CHALLENGER

Larry S. Bourgeouis, 30
Gary E. Coen, 34
Brock R. (Randy) Stone, 32
No women known to hold position during the Shuttle Program

(Ref. 48)

For completion of medical objectives and two payload medical experiments on STS-44, the Flight Surgeon Team got to hang the mission plaque. Shown here (left to right) are Kevin MacNeill (research support), Kristen Maidlow (biomedical engineer), John Charles (investigator), Roger Billica (flight surgeon), Robert Janney (biomedical engineer), and Richard Jennings (flight surgeon). On the far left is Flight Director Milt Heflin. To the far right is Astronaut Robert Cabana. NASA Photo S44-S-150.

FLIGHT SURGEON (SURGEON)

Shuttle flight surgeons were medical doctor whose patients were astronauts. They performed regular checkups, prescribed medications, and did research on the effects of space on their patients. They also had the authority to ground astronauts not found to be in proper physical shape before flight.

Before They Even Feel Bad

Dr. Roger Billica, MD, was branch chief of the Flight Surgeons in 1992. He said two surgeons were assigned per mission. "Those individuals start working with the crew at least six months prior to the flight," he said. "They do training and briefings as well as handle medical and health issues for the flight."

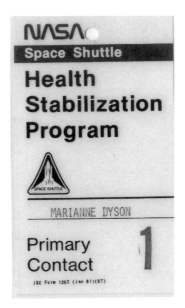

One of the first female astronauts, medical doctor Anna Fisher is shown here in 1978 participating in medical data collection used by the flight surgeons. NASA Photo S78-28265.

To keep the crew from being exposed to contagious diseases in the weeks prior to flight, only people who needed to interact with the crew in person, had passed a special physical, and had all the required vaccines were issued a Primary Contact badge that could be used to visit the crew in person. This badge was issued to the author because of the need to make last-minute changes to the checklists. Marianne Dyson photo.

A few days before launch, the flight surgeons packed their bags and followed the crew to Kennedy Space Center. "We spend a lot of time with the crew and their families prior to launch," Billica said. Then they staffed a console in the Launch Control Center during launch. "The crew surgeon is in (private) communication with the crew if need be," Billica said.

"As soon as they launch, we hop in a T-38, and fly lickity-split back here [Houston]." After a two-to-three-hour flight, they reported for duty in Mission Control.

Data from the onboard crew appeared on the flight surgeon's console in Mission Control. "We monitor the crew physical demands and physiological demands, and we listen to the conversations and the voices," Billica said.

Just like the pupils of the eyes adjust to bright light, parts of the human body adjust to the new environment of space. The flight surgeon studied each astronaut as they adapted, making sure the changes were, "normal."

Flight surgeons staff a position in the Firing Room at Kennedy Space Center, shown here in 1981, prior to launch. After launch, they flew back to Houston to monitor the crew's health from Mission Control. NASA KSC Photo 108-KSC-81PC-84.

Normal included getting a little dizzy. In space, the heart, used to pumping blood against gravity, pumps more than needed to the head. After a few days, it quits working as hard, creating the problem of fainting during entry. "We monitor their pulse and electrocardiogram in real time," Billica said. "We're watching that on console—we know what the metabolic [calorie] demands are based on exertion, the time, heart rate, and the breathing rate. All of those things are monitored to make sure we have a crewmember who is doing well and is also not getting in trouble from fatigue or strain on the heart or metabolic system."

Body fluids shift headward in space, making faces puffy and legs thinner. Without the pull of gravity, muscles weaken, and bones get brittle. The number of white blood cells also drops. "Space Adaptation Syndrome (SAS) is a broad range of things," Billica explained. "Space Motion Sickness (SMS) is one of those things. SMS to some degree occurs in about two thirds of crewmembers." (Ref. 2)

Unlike the early days when astronauts had to put up with feeling bad in space, flight surgeons developed an effective treatment for space sickness. "We found a medicine, phenergan, as an injection, works great," Billica said. "Once a crewmember gets symptoms, it's extremely effective in getting rid of them and not having them come back. We've been very pleased."

Billica said so far [in 1992], there was, "no essential difference between men and women" in terms of how they adapted to space. He cautioned however that, "we have a much larger data base of men than women. Add one more woman, and you have a significant change in the statistics. Right now, we do not

119

have a clear way to predict who is going to get it [space sickness]," Billica said. "If we did, we would do something about that."

Flight surgeons have found exercise prevents muscle weakness and slows the loss of calcium to the bones. Drinking extra fluids also helps prevent fainting during entry.

Flight surgeons rarely have to intervene in crew activities, but Billica recalled one time when they did. "The Lower Body Negative Pressure [LBNP] is a stress to the system to try and mimic gravity," Billica said. It is a kind of inflatable pants that force fluid back toward the head. "We have limits on how far we can go with this stress to the crewmembers," Billica explained.

"Our crewmembers are trained, they know when they are going to faint, and if they start getting symptoms, they would stop and open the seal on the LBNP."

But the flight surgeons don't wait for the crew to be in danger to act. "We can tell when they're heading in that direction. We stop it before they even feel bad," Billica said. "We called up and said, 'Stop that! Terminate LBNP.'"

We Plan on Being Ready

No medical emergencies occurred during any Space Shuttle flights. However, with longer flights, the flight surgeons expect to eventually see a broken bone in orbit. If that happens, "We plan on being ready," Billica said.

"We have two crewmembers trained to be medical officers. The Shuttle Orbiter Medical System has a checklist that tells them what to do," Billica said. "They wouldn't say, 'Oh my gosh, we've got a broken arm, let's call the Doc.' They know how to do the first response. They would then notify us and have a medical conference with the flight surgeon. That would be private—they would configure the Air/Ground so no one can overhear," Billica explained.

"We would consult with the crew medical officer and determine what the situation is and give them a treatment plan. Depending on the type of break, it may involve stabilizing it and continuing the mission or coming home," Billica said.

But there were no options for surgery onboard, even if one of the crewmembers were a doctor. The Shuttle lacked the necessary anesthesia and equipment. "We have IV solutions like you give someone in a field hospital, but we do not have blood replacement," Billica said. "Our level of capability to handle a true emergency is to stabilize and come home. There's a lot of things we deal with where we don't shorten the mission. But if it were a significant broken arm, we probably would talk about coming home."

The onboard medical kit shown here is used only for minor illness or injuries. The Flight Surgeons philosophy is to stabilize and return to Earth if anything more serious happens. NASA Photo S80-37864.

Flight Surgeons Richard Jennings, MD and Roger Billica, MD monitor the crew's adaptation to space from their console in Mission Control during STS-44 in 1991. To their right are PAOs Hal Stahl and James Hartsfield. NASA Photo S44(S)155.

It's a Very Serious Point

During preparations for return to Earth, the Flight Surgeons were busy too. "We work our shifts until the day before landing," Billica said. "Then we hop in a jet and fly to wherever the landing is going to be."

Many landings were delayed due to weather. When that happened, Billica said, "It's not uncommon for the flight surgeon to be up thirty-six to forty-eight hours with only catnaps." He said he would consider only having to stay awake twenty-four hours at a stretch, "a dream." Obviously, the ability to work well under pressure with no sleep was a must for flight surgeons.

Once they arrived at the landing site, they usually got a few hours of sleep though. "Then we're up several hours before landing getting ready. It's a busy time," Billica said.

"You've got a lot to do, a lot to coordinate and checkout. When the landing finally occurs, it's almost a feeling of relief. It's, 'Good! No emergencies!' They're down, and now it's just by the book."

Once the Orbiter came to a stop, a raised platform was wheeled over to the middeck hatch. "We actually go on the Orbiter when they roll up the white room," Billica said. "The doctors go in and assess the crewmembers and help them if they need it. Most of them don't need help, but we're there, and we stay with them for the next several hours." But sometimes the crew did need help.

"It's a very serious point," Billica said. "We're very professional, making sure they're okay. It's our responsibility."

In general, Billica said, "we have had to provide medical care at landing, mostly related to dehydration, nausea, vomiting, and feeling faint. In terms of balance and orientation, they are sometimes a little bit mixed up on returning to Earth. You do what's necessary to get them out."

After the doctors examined the crew and treated any problems, the crew exited the white room and

Using the raised platform seen here alongside the Orbiter after STS-26 landed in Edwards in 1988, the Flight Surgeons go onboard and help the crew out if they need it. In the photo, Vice President Bush greets the crew coming down the steps: Rick Hauck, Richard Covey, John (Mike) Lounge, George Nelson, and David Hilmers. NASA Photo S26-5168.

climbed down the steps. "Most of them do great," Billica said. "Then we're with them, kind of like a chaperon almost, making sure they're okay." After egress, there are usually several hours of medical tests.

"It's a demanding job, and it takes a lot out of you," Billica said. "When a flight is over, you want about a week or two of vacation!"

But in 1992, Billica didn't expect there to be much time for vacations as more flights were loaded with medical tests to prepare for the future. "Now that NASA has made a commitment for longer missions, for space station, and for exploration," Billica said, "there are a lot of life science issues that must be explored and resolved for us to successfully meet the long-term goals of NASA."

Billica remained the chief of the Medical Operations Branch at JSC from 1991 to 2001. For his work in developing resistive exercise as an effective countermeasure for bone loss in space, he was inducted into the Space Foundation's Space Technology Hall of Fame in 2019. (Ref. 39)

SHUTTLE SURGEON
STS-1 to STS-39

Name, first flight in MOCR

Sharon R. Tilton, 1 (first woman)
Michael A. Berry, 1
Michael W. Bungo, 2
Phillip C. Johnson, 2
James Vanderploeg, 2
Ellen L. Schulman, 3
James S. Logan, 3
Theodore E. Lefton, 6
P. M. Kuklinkski, 9
Donald F. Stewart, 51A
Sam Lee Pool, 51D (worked Apollo)
Jeffrey R. Davis, 51B
Patricia A. Santy, 61A

CHALLENGER
Richard Jennings, 26
John Schulz, 29
Phillip Stepaniak, 34
Larry Pepper, 32
Denise Baisden, 32
Brad Beck, 35
(next was Roger Billica, 44, quoted in chapter)

Note, this list may be incomplete. No data was found for STS-51C, 51I, 51J, 61B, 51L, 27, 28, and 33. (Ref. 48)

The failure of the original plan to bring Intelsat, shown here during rendezvous, into the payload bay for repair brought a lot of negative press attention to STS-49 in 1992. But PAO Jeff Carr knew the Flight Control Team would find a solution— and they did. NASA Photo S49-52009.

PUBLIC AFFAIRS OFFICER (PAO)

The Shuttle Public Affairs Officer was the voice of Mission Control. And for people not familiar with NASA jargon, PAO was also the translator.

The Public Affairs Officer was responsible for publicity before and after flights. They arranged astronaut interviews and press briefings. They prepared press kits, photos, brochures, press releases, speeches, and answered questions from reporters. To stay on top of mission events, they also spent a lot of time in simulations in Mission Control.

STS-49 INTELSAT really brought public concern during the
EVAs. Just to give you a tickle for the day, read along and
see all the ideas that were suggested.

1. The crew must have magnets onboard. Surely they can put
    the magnets on their feet, hold their feet in the air and
    therefore be able to pull the satellite to the shuttle.

2. Those boys obviously are't from Texas, otherwise they
    would be able to throw a lasso around the satellite
    and draw it in.

3. Take two tethers, throw one around each end of the
    satellite, hook the other ends to the shuttle and reel it
    in.

4. Throw a rope around the satellite and use it as a noose,
    then have a stabilizing bar to hold the satellite still,
    then draw it in.

5. The question was asked if they would spin up the
    satellite this evening like they did last night. When
    they were told yes the man replied, "I don't think that
    will help the astronauts capture the satellite, and I
    personally wish NASA would not do that. In my opinion
    that would not be the right thing to do", and then he
    hung up.

6. A gentleman from Massachusetts suggest that we use two
    large fishing hook, twirl them over our head, latch the
    hooks to the satellite and then reel it in.

7. Use velcro, if not velcro then some other sticky
    substance.

8. Use a rod with a hook on the end to grip one side of the
    satellite, then lasso the other side. In theory (to this
    person) you would be able to stabilize the satelite with
    the rod while reeling it in via the tether.

9. Bungee cords (I never did understand how this man thought
    they would work).

10. The last gentlman that called at 11:15 suggested that
    Rick Heib hold Pierre Thuot and spin Pierre at the same
    rate of speed as the Satellite and then attach the bar.

STS-49 Rescue Ideas Sent to PAO

*Dealing with the public was the job of the PAO. Here are some of the ideas the public sent for ways
to capture the Intelsat satellite during STS-49. (Ref. 5)*

The Vultures Were Gathering

In 1992, Jeffrey E. Carr was a Public Affairs Officer during STS-49 which involved the capture and repair of the Intelsat satellite. "Attention to the spacewalk during the first attempt at capture with a grapple bar was kind of lukewarm," Carr remembered. "It got lip service on most of the network news programs and in most of the papers, but it was just another spacewalk and another satellite rescue and kind of ops normal in the minds of a lot of people," Carr said. But all that soon changed.

"When the capture bar didn't work, out came the vultures!" Carr recalled. "Suddenly, the news center over in Building 2 was doubly populated. There were correspondents coming in from Dallas and Atlanta to zero in on what was certain to be another NASA failure, another gloom and doom story. They were already writing the obits [obituaries] for us and gathering for the kill." But Carr wasn't concerned.

"We could sense there was apprehension going into the second EVA, but the control center is isolated from that kind of stuff. We just sort of watched and held our breath," Carr said. The second attempt to capture the satellite also failed.

"After two failed attempts to capture the satellite using methods and tools originally designed for the job, the Flight Control Team had to improvise," Carr said. "The people on the control team weren't worried about the headlines or the vultures gathering at the door. They were upbeat and positive. They looked at what had happened on the second EVA and said, 'well this capture bar is not going to work—not because it's a bad capture bar but because we underestimated the mass dynamics of this big satellite.'"

They decided to try a never before attempted three-man EVA and grab the satellite with their hands. "The flight controllers were taking a positive approach to solving the problem, and I knew, I could tell from the environment in the control center, that they were going to solve the problem," Carr said.

"That's when I began to enjoy my perch: my perspective on both sides of the fence—watching what was going on inside and watching the ambulance chasers out there. We were going to put the EVA off another day giving them time to stew and speculate and sensationalize. On the other hand, I knew we were going to solve the problem, so I was beginning to get very excited. What I sensed was just the opposite of what the news media sensed. I knew we were going to beat them.

"Public interest was very high, and all the major national news organizations covered the third spacewalk," Carr said. "It was a lot of fun watching that EVA when the three astronauts went outside and grabbed the satellite with their hands. That was probably one of the greatest triumphs over public expectation that we had in the Shuttle Program that I can remember," Carr said.

Intelsat was captured, repaired, and put back into service. A success story all around. "Describing the events as they unfolded—knowing history was being made—was extremely satisfying," Carr said. (Ref. 5)

A Record-Breaking Flight

The next flight, STS-50 (June 1992) was another big success for NASA and also a personal success for Carr.

"It was two days prior to launch," Carr recalled, "and we were launching under our new Associate Administrator for Public Affairs (Livingston) who was a very forward-thinking, very aggressive leader. He

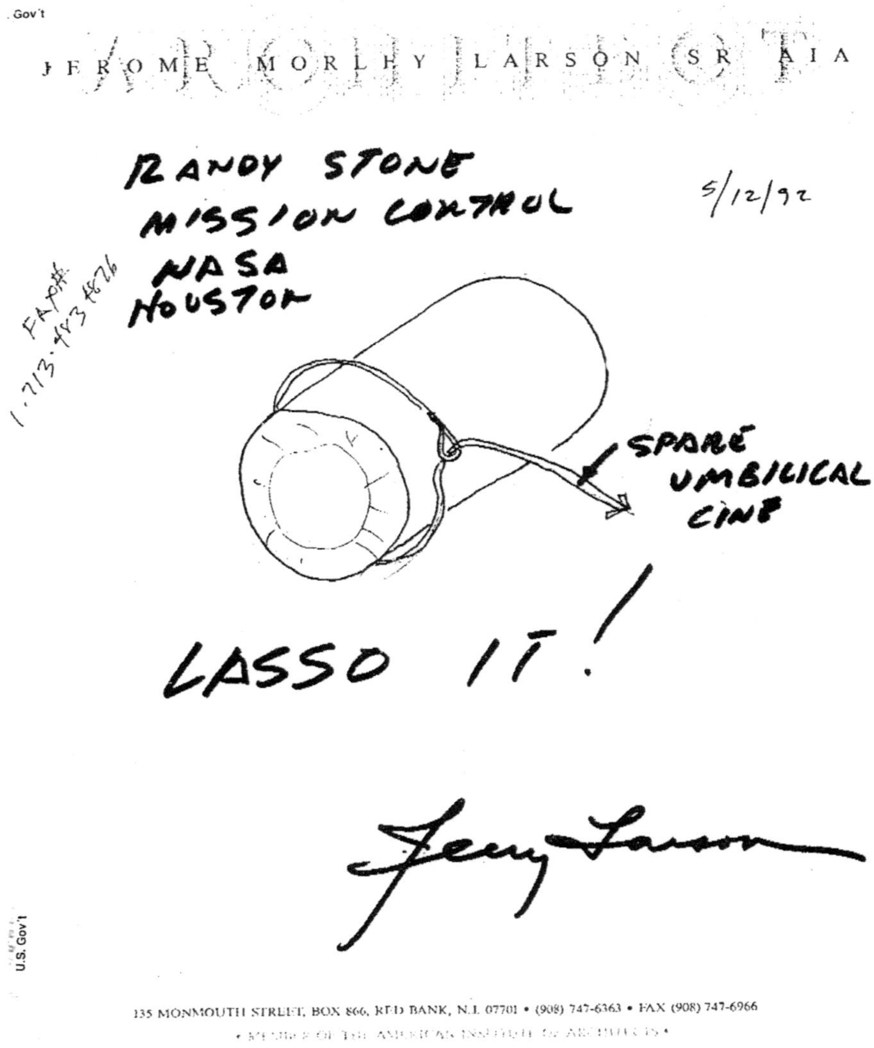

RANDY STONE
MISSION CONTROL
NASA
HOUSTON

5/12/92

FAX# 1-713-483-4876

SPARE
UMBILICAL
CINE

LASSO IT!

Jerry Larson

U.S. Gov't

135 MONMOUTH STREET, BOX 866, RED BANK, N.J. 07701 • (908) 747-6363 • FAX (908) 747-6966
• MEMBER OF THE AMERICAN INSTITUTE OF ARCHITECTS •

Jerome Larson sent a fax to Flight Director Randy Stone in 1992 saying that "Those boys obviously aren't from Texas, otherwise they would be able to throw a lasso around the satellite and draw it in." None of the STS-49 crewmembers were from Texas. The original two spacewalkers, Rick Hieb and Tom Akers were from North Dakota and Missouri, respectively. Address and phone numbers cropped by the author.

informed me we had an urgent situation. Amendments were being introduced on the floor of the House in Congress that were designed to kill funding for space station. It was imperative in his view that we do everything we could to get the word out through the news media that the things we were doing on this mission were significant groundwork for space station."

New equipment on this flight allowed it to last a record of thirteen days, nineteen hours, and thirty minutes. "This would demonstrate we were already making vast commitments to Freedom [Space Station which evolved into the International Space Station]. The kinds of things we were doing would stand to better all of our lives," Carr said.

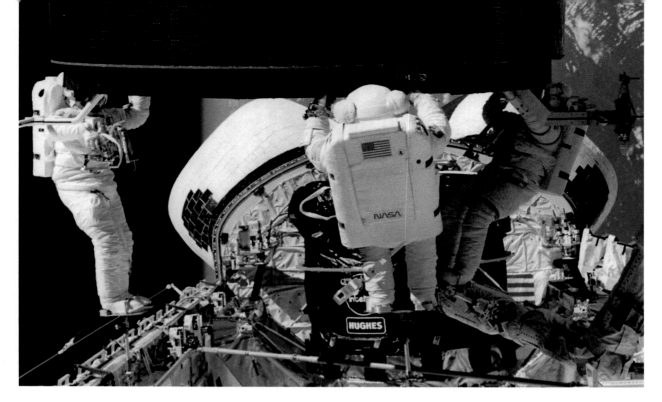

The media "vultures" were swooping, but PAO Jeff Carr rightly predicted the third EVA would be the charm on STS-49. L to R: Astronauts Rick Hieb, Thomas Akers, and Pierre Thuot. NASA Photo S49-16014.

"With very little time and knowing the rigidity and structure of the Shuttle missions, we had our work cut out for us. Livingston wanted us to arrange, on an every-day basis, inflight interactive broadcast events to place the crew live on major international news programs." This would be a challenge at any time, but it was especially so two days before launch.

"We held an emergency meeting with the lead Flight Director, the lead FAO, the lead Payloads Officer, and our television production staff," Carr said. Within a day, we had combed the timeline and found at least one candidate opportunity each day. We had to do some minor rescheduling, but we were successful without impacting any major science operations that had been in planning for years.

"With these opportunities in place, my job was then to negotiate these press appearances with national network news programs such as Good Morning America, CBS This Morning, NBC Today, as well as network radio such as Mutual Radio and CBS Radio," Carr explained.

Not all the events were scheduled by launch day. "Throughout the flight, we were constantly negotiating events with the networks, planning who would provide audio and voice circuits between the network and Houston," Carr said.

"Then, prior to the event, with the actual network talent who would be conducting the interviews, we would conduct an end-to-end voice check from their studio all the way to the comm tech which is the last point before the voice on the ground gets into the air/ground voice system to the crew. We didn't have time to do checkouts end-to-end with the crewmembers on board because of their limited availability, so we had to do everything we could on the ground to assure we had a link," Carr said.

During the week of the Space Station vote in the House in 1992, PAO Jeffrey Carr scheduled eight in-flight press conferences with the STS-50 crew. From top, L to R) Kenneth D. Bowersox, Lawrence J. Delucas, Richard N. Richards, Bonnie J. Dunbar, Carl J. Meade, Ellen S. Baker, and Eugene H. Trinh. MSFC-9262230, 6-5-1992.

They also tried to ease the job of the interviewers. "You never know what degree of expertise the correspondent—the talent at the other end—is going to have," Carr explained. "We were doing everything we could to pull together information packets and faxing them out to the correspondents hours before the event. We learned that correspondents are not offended by being coached in advance of interviews, but grateful to have as much information as they can about what's important about the flight."

The broadcast events were a big hit. "In the course of five flight days, we conducted eight inflight broadcast events," Carr recalled. "Normally, there are only two."

Despite having only two days to prepare, the Public Affairs Team got the job done. "Everything went like clockwork, it really did," Carr said. "We knew our business, we were in control, and the networks did not have the luxury of dictating to us. Because of that, we felt confident, and things went without a hitch."

As they say, necessity is the mother of invention. "Under the gun, we learned a lot about streamlining these events and finding ways to do them easier, cheaper, faster, and more reliably," Carr said.

"The feedback was outstanding. We found that the exchanges and dialogue from the one-on-one interviews produced significant news coverage, more substantial coverage than any flight press conference ever done." But, perhaps even more important, the extra press coverage of another successful flight came at a crucial time for NASA.

PAO Jeffrey Carr moderates a press briefing with STS-50 Flight Director Randy Stone. For his efforts getting public attention for space during STS-50, Carr became the first PAO to hang the mission plaque. NASA Photo STS-50-077.

PAO Janet Ross, shown here during STS-51G in 1985, was one of the first women to be the "Voice of Mission Control." The Gemini, Apollo, and Skylab Programs all had male Public Affairs Officers. NASA photo 51g-s-213.

"The vote came about two thirds the way into the flight and the resolution to delete space station funding was soundly defeated," Carr said with relief.

The determined Public Affairs Officer's efforts were not overlooked by the Flight Control Team. "Based on the extremely positive coverage in the national and regional markets," Carr said, "the Flight Director felt we had made the most substantial contribution to the success of the flight and awarded us the honor of hanging the plaque." This was the first time that a public affairs officer was given this honor. (Ref. 5)

Carr was the "Voice of Mission Control" for more than 40 Space Shuttle missions. He because director of Public Affairs and news chief at JSC and served as a special assistant to the NASA Administrator. He left NASA in 1996 and became the director of Communications and Public Relations at United Space Alliance. In 2010, he joined Griffin Communications Group and became its president in 2014. (Ref. 44)

SHUTTLE PAO
STS-1 to STS-39

Name, first flight in MOCR

John E. McLeaish, 1
Harold (Hal) Stahl, 1
L. John Lawrence, 1

[No data for STS-2 through 51-B]

R. Terry White, 51G
Janet K. Ross, ?51G (?first woman)
Billie A. Deason, ?51G (?first woman)
Brian Welch, 51F
Steve Nesbitt, 51F

CHALLENGER
Barbara L. Schwartz, 26
Jeffrey E. Carr, 29
Kyle J. Herring, 34
Kari L. Fluegal, 41
James A. Hartsfield, 35
Pam Alloway, 37

Note, this list is incomplete. Data is missing for STS-2 through 51G, and STS-51C, 51A, 51J, and 61B. (Ref. 48)

BIBLIOGRAPHY

1. Adams, Bob. Questionnaire interview response. April 2, 1992, and review comments, Jan. 7, 1993.

2. AOPA Handbook for Pilots. p 219. AOPA Air Safety Foundation, 421 Aviation Way, Frederick, MD 21701. 1993.

3. Billica, Roger, MD Personal interview. Taped October 9, 1992.

4. Brekke, Michele. Questionnaire interview response. April 2, 1992. Personal interview. Taped October 8, 1992. Review comments received November 3, 1992.

5. Carr, Jeff. Questionnaire interview response. July 14, 1992. Personal interview. Taped October 16, 1992.

6. Castle, Bob. Questionnaire interview response. February 24, 1992. Review comments received November 5, 1992.

7. Conley, Carolynn. Personal interview. Taped October 6, 1992.

8. Darnell, Michael. Questionnaire interview response. April 20, 1992.

9. Dye, Paul. Questionnaire interview response. February 24, 1992. Review comments received November 3, 1992.

10. Foy, Larry. Personal interview. Taped September 29, 1992. Review comments received October 26, 1992.

11. Fugitt, Mark D. Questionnaire interview response. March 11, 1992. Review comments received November 17, 1992.

12. FRC1-12 Data Sheet. Lockheed Missiles and Space Company. PO Box 3504, Sunnyvale, CA 94088. Dated January 1987.

13. Giannukos, Linda. Photos of General-Purpose Computers. Communications Secretary, IBM Federal Systems Company, 3700 Bay Area Boulevard, Houston, TX 77058. October 16, 1992.

14. Ham, Linda. Personal interview. Taped December 31, 1991.

15. Harwood, William, "NASA's State-of-the-Art Control Center to Debut in June," Space News, May 22-29, 1995.

16. IBM Federal Sector Division. Shuttle Computer Systems Press Information. International Business Machines Corp., 3700 Bay Area Boulevard, Houston, TX 77058. February 1991.

17. Information from a letter to Mr. James Fletcher, Administrator of NASA, from Stan Graves, State Historic Preservation Officer, Texas Historical Commission, regarding the designation of the Mission Control Center as a National Historic Landmark. September 9, 1987.

18. Knight, Jack. Questionnaire interview response. March 24, 1992. Review comments received November 2, 1992.

19. Kranz, Eugene. Personal interview. Taped April 19, 1991 and March 9, 1992. Review comments received November 16, 1992.

20. Loree, Ray. MCC Development History. Unpublished. Houston, TX. August 1990. Based on conversations and comments from NASA and Contractor co-workers; an article, "The Evolution of the Mission Control Center," by Mike Kearney published in the Proceedings of the IEEE, Volume 75, No. 3, March 1987; and from the FAC Administration Plan, Attachment F, January 10, 1986.

21. Markle, Frank. Personal interview. Taped September 16, 1992. Review comments received November 4, 1992.

22. McDede, Jim. Personal interview. Taped September 30, 1992. Review comments November 17, 1992.

23. NASA. Johnson Space Center. Houston, TX. Public Affairs Office. Press Releases, "Flight Control of -," individually released for missions from Gemini 4 through STS-44.

24. NASA. Johnson Space Center. *Flight Control Operations Handbook*. Volume I, Shuttle OPS, STS. Mission Operations Directorate. Facility and Support Systems Division. Houston, TX. Final, Rev C, PCN-1, JSC-12805, April 1, 1989.

25. NASA. *The Gamma Ray Observatory*. NASA document number NP-124. Received February 25, 1991.

26. NASA. Lyndon B. Johnson Space Center. *Mission Operations Directorate Orientation Manual*. Mission Operations Directorate. Training Division. Houston, TX. JSC-20349.

27. NASA. Johnson Space Center. *PDRS Console Handbook*. Volume I, Basic, Rev. A. PDRS Responsibilities, Section 1. Originator: R. Zaguli. Mission Operations Directorate. Facility and Support Systems Division. Houston, TX. JSC-17525, December 1, 1989.

28. NASA. US Government Printing Office. *Shuttle News Reference*. 1981. 723-000/562.

29. NASA. Johnson Space Center. Houston, TX. Public Affairs Office. "Replacing the irreplaceable." *Space News Roundup*. Volume 29 #1. January 5, 1990.

30. Perry, Brian D., Questionnaire interview response. March and June 1992. November 3, 1992.

31. Presley, Willard. Personal interview. Taped October 6, 1992.

32. Rockwell International. Rocketdyne Division. 6633 Canoga Ave., Canoga Park, CA 91304. *Shuttle Main Engine incredible facts*, and photos of the main engines. Contact Kristy Varnes. Pub. 574-N-13. Rev. 8-91.

33. Thiokol Corporation Fact Sheet. Thiokol Communications Office, 2475 Washington Blvd., Ogden, UT 84401. Dated October 7, 1992.

34. Tour guide speeches about Mission Control. Taped Nov. 1991.

35. Trlica, Ed, Jr. Personal interview. Taped October 7, 1992, and comments Jan. 7, 1993.

36. United Technologies Hamilton Standard News. "Space Shuttle Space Suit/Life Support Systems." Press release for STS-37. Contact John Mayo (203-654-4790), Windsor Locks, Connecticut 06096. Received February 25, 1991.

37. Walz, Carl E. Personal interview. Taped October 20, 1992.

38. Zaguli, Ron J. Personal interview. Taped March 2, 1992. Received comments November 3, 1992.

References added in 2021 (not in alphabetical order with the original bibliography above)

39. Andrews, M. "SpiraFlex Interim Resistive Exercise Device (iRED)." 4-17-19. Accessed 2-20-21. Available online: https://www.spacefoundation.org/space_technology_hal/spiraflex-interim-resistive-exercise-device-ired/

40. Ancestry. "Williard Presley 1941-1997." Accessed 2-20-21. Available online: https://www.weremember.com/willard-presley/1i6i/memories?utm_campaign=guestsWelcome&utm_source=email

41. Dyson, Marianne. *A Passion for Space: Adventures of a Pioneering Female NASA Flight Controller.* Springer, 2015. p. 298.

42. Dyson, Marianne. *Space and Astronomy: Decade by Decade.* Facts on File, 2003. p 198-99.

43. Jordan, Gary. "How the Mission is Controlled." NASA JSC. Dec. 19, 2019. Downloaded 1-18-21. Available online: https://www.nasa.gov/feature/how-the-mission-is-controlled-inside-nasa-and-boeing-joint-operations

44. Linkedin. "Jeff Carr." Accessed 2-20-21. Available online: https://www.linkedin.com/in/jeff-carr-2b54b57/

45. Linkedin. "Darnell, Michael." Accessed 3-2-21. Available online: https://www. Linkedin.com/in/michael-darnell-4142a116/

46. Linkedin. "Mark Fugitt." Accessed 2-20-21. Available online: https://www.linkedin.com/in/mark-fugitt-b366157/

47. Manned Spaceflight Operations Association. "Manning Lists." Last Update 01-20-21. Available online. https://www.mannedspaceops.org/about-us/manning-lists/

48. NASA. "A SAFER Way to Space Walk." 2-27-2004. Accessed 2-20-21. Available online: https://www.nasa.gov/missions/shuttle/f_saferspacewalk.html

49. NASA Johnson Space Center. *Space News Roundup.* September 21, 1984. Accessed 2-8-21. Available online: https://historycollection.jsc.nasa.gov/JSCHistoryPortal/history/roundups/all_roundups.htm

50. NASA Johnson Space Center. "Veteran Astronaut Carl Walz Leaves NASA." Dec. 04, 2008. Accessed 2-24-21. Available online: https://www.nasa.gov/home/hqnews/2008/dec/HQ_08-318_Walz_Leaves.html.

51. NASA. National Space Transportation System Reference. June 1988.

52. Pleora Technologies. "In-orbit Inspection." Accessed 2-8-21. Available online: https://www.pleora.com/markets/transportation/in-orbit-inspection/.

53. Regents of the University of Minnesota. News Release, "Michele A. Brekke to be University of Minnesota Twin Cities campus 2019 Homecoming Grand Marshal." August 23, 2019. Accessed 2-20-21. Available online: https://twin-cities.umn.edu/news-events/michele-brekke-be-university-minnesota-twin-cities-campus-2019-homecoming-grand-marshal

54. Spaceflight Now. William Harwood. "NASA space telescope heads for fiery crash into Pacific." May 28, 2000. Accessed 3-5-21. Available online. https://en.wikipedia.org/wiki/Compton_Gamma_Ray_Observatory

55. Youtube.com. "STS-51-L Mission Controlling in 1986 (Challenger Space…)" Accessed 2-20-21. Available online: https://video.search.yahoo.com/search/video?fr=mcafee&p=youtube+launch+of+Challenger+51L#id=53&vid=85af1e2f2e4356c18fec651559da3c3d&action=view]

ACRONYMS AND ABBREVIATIONS

APU: auxiliary power unit

ATO: Abort-To-Orbit

BFS: backup flight system

CAPCOM: capsule communicator, astronaut in Mission Control

DoD: Department of Defense

DPS: data processing system

EECOM: environmental consumables and mechanical engineer (MCC position)

EGIL: electrical generation, instrumentation, and lighting engineer (MCC position)

EVA: extravehicular activity, i.e. spacewalk

FAO: flight activities officer (MCC position)

FCR: flight control room, FCR-1 is on the 2nd floor, and FCR-2 on the third floor of Building 30

FD: flight day on a timeline, Flight Director on console button

FDO: flight dynamics officer (MCC position)

FLIGHT: Flight Director (MCC position)

FOD: Flight Operations Directorate and MCC position

GC: ground controller

GUIDANCE: guidance officer (MCC position)

GNC: guidance, navigation, and control (MCC position)

GPC: general purpose computer

GPO: guidance and procedures officer (MCC position)

IFM: in-flight maintenance

IMU: inertial measurement unit

INCO: instrumentation and communications officer (MCC position)

IPS: instrument pointing system

ISS: International Space Station

JPL: Jet Propulsion Lab

JSC: Johnson Space Center

KSC: Kennedy Space Center

LEM: Lunar Excursion Module

MCC: Mission Control Center

MECO: main engine cutoff

MMACS: maintenance, mechanical, arm, and crew Systems (MCC position)

MOC: mission operations computer

MOCR: mission operations control room

MOD: Mission Operations Director (MCC position) and Mission Operations Directorate

MPSR: multi-purpose support room, a.k.a. "back" room or SSR

MSFC: Marshall Spaceflight Center

OEX/DAP: orbiter experiments digital autopilot

OMS: orbital maneuvering system (engines)

PAM: payload assist module

PAO: Public Affairs Office or officer (MCC position)

PDRS: payload deployment and retrieval system

POCC: Payloads Operations Control Center

PROP: propulsion engineer (MCC position)

RCS: reaction control system (thrusters)

RMS: remote manipulator system, also MCC position

RMU: remote manipulator systems, mechanical systems, and upper stages (MCC position)

RNDZ: rendezvous

RSO: range safety officer

SAS: space adaptation syndrome

SMS: Shuttle Mission Simulator and space motion sickness

SSR: system support room, a.k.a., "back" room

STS: space transportation system

TDRS: tracking data relay satellite

TV: television

WCS: waste control system

ZOE: zone of (communications) exclusion

APPENDIX A

Shuttle Flight Control Individuals/Teams Honored to Hang the Mission Plaque

Flight	Launch Date	Position/Flight Controller
ALT-1	08/12/77	FDO/Chuck Deiterich
STS-1	04/12/81	EECOM/Jack Knight
STS-2	11/12/81	EGIL/Paul Joyce
STS-3	03/30/82	INCO/Lee Briscoe
STS-4	06/27/82	PROP/Ron Dittemore
STS-5	11/11/82	GC/George Egan
STS-6	04/04/83	DPS/Mike Darnell
STS-7	06/18/83	WEATHER/Jim Nicholson
STS-8	08/30/83	FAO/Cheevon (Mi-Mi) Lau
STS-9	11/28/83	INCO/Bob Castle & FAO/Bob Nute
STS-41B	02/03/84	EVA/EVA Team, Terry Neal - Lead, Barry Boswell, Bill Veach (hung plaque).
STS-41C	04/06/84	RENDEZVOUS/Rick Hieb
STS-41DR	08/30/84	RMS/Ron Zaguli & Mark Ferring
STS-41G	10/05/84	PAYLOADS/Michelle Brekke & Linda Godwin
STS-51A	11/08/84	FDO/Brian Jones
STS-51C	01/24/85	OIO/Kim Anson
STS-51D	04/12/85	EVA/Charlie Armstrong & IFM/Robbie Robbins
STS-51B	04/29/85	NSTS Program Manager/Glynn Lunney
STS-51G	06/17/85	PAYLOADS/Michelle Brekke
STS-51F	07/29/85	FDO/Brian Perry & BOOSTER/Jenny Howard (now Stein)
STS-51I	08/27/85	FDO/Brian Jones
STS-51J	10/03/85	OIO/Kim Anson (hung plaque) & PAYLOADS/Jim Simons
STS-61A	10/30/85	PROP/Rich Jackson
STS-61B	11/26/85	PROP/Linda Hautzinger (later Ham) & Jim McDede (hung plaque)
STS-61C	01/12/86	FAO/Phil Engelauf
STS-51L	01/28/86	Plaque hung in FCR 2 (historical monument) on the 17th anniversary of the *Challenger* launch in a small ceremony. Lead FD Randy Stone hung the plaque near the Apollo 1 plaque where it will stay until the completion of the mission. Randy Stone stated, "these men and women are still flying."
STS-26	09/29/88	EECOM/Steve McClendon
STS-27	12/02/88	MER THERMAL/Bobby Brown, Luther Palmer, Harry Chang, Don McCormack, Tommy Taylor, Joel Krouse
STS-29	03/13/89	PAYLOADS/Ben Sellari

STS-30	05/04/89	DPS/Terry Keeler & IFM/Paul Lloyd
STS-28	08/08/89	FAO/Karen Engelauf
STS-34	10/18/89	WEATHER/Dan G. Bellue, Scott J. Cunningham, Gene M. Hafele, Dennis J. Haller, Edward K. Hogan, Charles C. Morrill, Doris A. Rotzoll & Steven J. Sokol (hung plaque).
STS-33	11/22/89	FLT DES Team - Lead/John Fields (hung plaque) & FDO Team - Lead/Ed Gonzalez & PROP Team - Lead/Karen Jackson
STS-32	01/09/90	FDO/Mark Haynes (hung plaque) & FLT DES -Lynda Slifer (now Gavin), Richard Gavin
STS-36	02/28/90	FAO/Gail Schneider (John Walsh hung plaque)
STS-31	04/24/90	PAYLOADS/Nellie Carr (hung plaque), Jeff Hanley, Cheryl Molnar, Jeff Larson, Pete Sprunger
STS-41	10/06/90	GC/Larry Foy
STS-38	11/15/90	PAYLOADS/Mark Childress (hung plaque), FLT DES/Sally Davis, FAO/Pete Hasbrook.
STS-35	12/02/90	GNC/Will Fenner (hung plaque), Ken Bain, Ed Trlica, Linda Patterson, Bob Ellsworth, Kevin Dunn, Jeff Wyrick
STS-37	04/05/91	EVA/Bob Adams (hung plaque) & EVA Team
STS-39	04/28/91	PAYLOADS/Mark Kirasich & FAO/Neil Woodbury & PAYLOADS & FAO Teams.
STS-40	06/05/91	PAYLOADS/Roger Galpin & Payload Team.
STS-43	08/02/91	TRAINING/J.P. Poffinbarger (hung plaque) & WEATHER/Charles Morrill & Training & Weather Teams.
STS-48	09/12/91	FDO/Steve Stich & EGIL/Bob Armstrong
STS-44	11/24/91	SURGEON-Medical Operations Branch/Roger Billica
STS-42	01/22/92	EECOM/Dan Molina & PROP/Matt Barry and PROP and EECOM Teams.
STS-45	03/24/92	INCO/Chris Counts
STS-49	05/07/92	FDO/Mark Haynes, RNDV/Chris Meyer & EVA/Jerry Miller
STS-50	06/25/92	PAO/Jeff Carr, Kyle Herring, Mike Curie (Hung Plaque), Diana Ormsbee, Janet Packham
STS-46	07/31/92	PL/Jeff Hanley, GPO/John Malarkey, and Payload and Dynamics Teams.
STS-47	09/12/92	MMACS/Kevin McCluney and the MMACS Team
STS-52	10/22/92	FAO/Tony Griffith (hung plaque), PL/Gene Cook and the FAO and PL Teams.
STS-53	12/02/92	Security/Mike Corbin (hung plaque) & Carol Smith.
STS-54	01/13/93	EGIL/Bob Armstrong and Steve Hirshorn
STS-56	04/08/93	INCO/Chris Counts, Harry Black, Joe Fanelli, J. Conner
STS-55	04/26/93	GC/Lynn Vernon & the ground communications team.
STS-57	06/21/93	FAO/Greg Smith and the FAO team.
STS-51	09/12/93	PL/Tim Baum (hung plaque), Yvette Shannon & the PL Team
STS-58	10/18/93	EECOM/Quinn Carelock and the EECOM Team.
STS-61	12/02/93	EVA/Jim Thornton, Sue Rainwater, and the EVA Team.
STS-60	02/03/94	FAO/John Curry, John Tolle, Tracy Calhoun, Stacie Greene, Chris McKenna (hung plaque) and the FAO Team.
STS-62	03/04/94	MMACS/Bill Anderson, and MER/Dave Young.

STS-59	04/09/94	Pointing/Larry Friedl & IFM/Mike Maher
STS-65	07/08/94	Navigator/Jim Lassa
STS-64	09/09/94	Weather/Steve Sokol (Hung plaque) & DPS/Kim Terry
STS-68	09/30/94	FDO/Roger Simpson and Flight Dynamics Team
STS-66	11/03/94	FAO/Rich Cilento and FAO Team
STS-63	02/03/95	Prop/Robbie Gest & Rendezvous/Joe Williams
STS-67	03/02/95	Weather/Tim Garner, Karl Silverman, Mark Keehn, Richard LaFosse, Steve Sokol, Doris Rotzoll, Wayne Baggett, Brian Batson, Dan Bellue, Monica Sowell, and Pavlina Papado-poulous & IPS/Jeff Wyrick, Terry Morphy, Terry Keeler, Kevin Dunn, Eddie Trlica, Kathy Akag, Ken Bain and Michelle LaFleur.
STS-71	06/27/95	RNDZ/Lynda Gavin (hung plaque), MMACS/Karl Pohl, FAO/John Curry, FLT DES/Joel Montalbano, Integration/Greg Lange.
STS-70	07/13/95	GC/Chuck Capps, Wx/Steve Sokol and GC Team (Larry Foy, Norm Talbot, Joe Aquino, George Egan, Melissa Blizzard, Mike Marsh, John Snider, John Wells, Jonnie Brothers)
STS-69	09/07/95	PL/Jeff Hanley (hung plaque) & GNC/Jeff Davis and Payloads and GNC teams.
STS-73	10/20/95	EECOM/Jim Spivey & GC/Henry Allen
STS-74	11/12/95	FAO/Gail Schneider and FAO team.
STS-72	01/11/96	PAYLOADS/Susan C. Beisert (hung plaque), Roger A. Galpin, Helen E. Dutton, Jefferey A. Larson, Lee M.Gonzales, Amie Allison, Jim Runke, Mark T. Severance, Robert V. Grilli, Donna E. Stephenson, Don R. Settle, Susan A. Horelka & Thermal/Ray Serna, Rick Miller, Cathy Rose
STS-75	02/22/96	PLO/Jeff Hanley (hung plaque) and RGPO/Joe Williams, GC/Terry Quick and Training/ Bob Mahoney representing the Payloads, Tether Dynamics, Ground Control/Tracking Station and Training teams.
STS-76	03/22/96	EECOM/Dan Molina
STS-77	05/19/96	RNDZ/Dave Harshman & Bryan Bertrand
STS-78	06/20/96	IFM/Jeff Stone
STS-79	09/16/96	Prop/Bryan Lunney and MMACS/Kevin McCluney
STS-80	11/19/96	Lead FDO/Bill Britz and Ascent FDO/Matt Abbott
STS-81	01/12/97	GNC/Ken Bain (hung plaque), Kevin Dunn, Eddie Trlica, Laura Stallard, Dave Marquette, Mike Sarafin and Jeff Wyrick of the GNC team.
STS-82	02/11/97	EVA/Oscar Koehler and the EVA team.
STS-83	04/04/97	EGIL/Ray Meissler and the Fuel Cell Tiger team.
STS-84	05/15/97	FAO/Greg Smith, Mike Schieb(hung plaque) and the Lead FAO team. Joel Montalbano also noted for dual work as RIO and Flight Design Manager
STS-94	07/01/97	Ray Barrington (EG/Draper Labs), Pointing/Scott Patano (hung plaque), and GNC/Kevin Dunn
STS-85	08/07/97	Payloads/Rob Napp & Amie Allison
STS-86	09/25/97	Payloads/Yvette Shannon & FAO/Roger Smith
STS-87	11/19/97	Prop/Cathy Larson (now Koerner) & the PROP Team
STS-89	01/22/98	FDO/Bill Tracy (hung plaque) and EECOM/Jimmy Spivey

STS-90	04/17/98	IFM/Victor Badillo and EECOM/Steve Koerner
STS-91	06/02/98	DPS/Terri Murphy, Robert Hudson (hung plaque) and the DPS team.
STS-95	10/29/98	FAO/Roger Smith, Terri Schneider (Lead Timeline) and Kieth Lawson (Lead Pointer, hung the plaque).
STS-88	12/04/98	MMACS/Kevin McCluney, PDRS/Angela Prince & EVA/Scott Bleisath (hung plaque).
STS-96	05/27/99	ACO-Transfers/Ursula Stockdale (hung plaque) & Moscow FD/Dimitri Kovrichkin representing all Moscow FDs (Kolchin, Budnik etc.) Honorable mention.
STS-93	07/22/99	Payloads/David Brady and Booster/Jon Reding
STS-103	12/19/99	Lead Timeline/Doug Bristol and Payloads/John McCullough for the Payloads and FAO teams
STS-99	02/11/00	Prop/Dean Lenort and the STS-99 PROP Team.
STS-101	05/19/00	FDO/Dan Adamo
STS-106	09/08/00	Assembly and Checkout Officer/Rob Banfield.
STS-92	10/11/00	MMACS/Mel Friant (hung plaque), John, Shimp, Mark Welch and EGIL/Tim North (held ladder). Honorable Mentions: EVA/Daryl Schuck, Mike Hembree, Michelle Hollinger, Rndz/Jeannette Spehar, Prop/Cori Kerr, ACO/John McCullough, FAO Team/Greg Smith, Karen Watts and Neisha Lopez, RMS/Ian Mills, LSO/Marty Linde, Weather/Richard LaFosse.
STS-97	11/30/00	EVA/Glena Laws and John Haensly, Honorable mention – Training Team, Bill Todd, Delores Rader, Juan Garriga.
STS-98	02/07/01	EVA/Kerri Knotts and the EVA team and PDRS/Al Lee and the PDRS team.
STS-102	03/08/01	ACO/Jim Ruhnke and the ACO team.
STS-100	04/19/01	EVA/Jeff Patrick (hung Plaque)
STS-104	07/12/01	EVA/Oscar Koehler (hung plaque)
STS-105	08/10/01	GC/Bill Foster
STS-108	12/05/01	Lead Transfer Officer/Kristi M. Duplichen
STS-109	03/01/02	EVA/Dana Weigel
STS-110	04/08/02	EECOM/Paul Bolton and FDO/Chris Edelin honorable mention.
STS-111	06/07/02	GNC/Michael Sarafin
STS-112	10/07/02	Sheri Gray (hung plaque) of the CC Power down and Recovery Team.
STS-113	11/23/02	Cori Kerr (hung plaque), Lead PROP, on behalf of the entire PROP team, for outstanding consumables management.
STS-107	01/16/03	xxx
STS-114	08/09/05	PDRS/Michael Wright (hung plaque), with Team 4 FD/Kelly Beck holding the ladder
STS-121	07/17/06	The FAO Team hung the STS-121 Mission Plaque (Terry Clancy, Jaime Marshik and Laura Eadie). Jaime Marshik hung the plaque with Rick Lafosse (Spaceflight Meteorology Team) holding the ladder.
STS-115	09/21/06	The STS-115 Mission Plaque was hung by Lead INCO/Douglas Branham and PDRS/Scott Wenger held the ladder.
STS-116	01/22/07	The Spaceflight Meteorology Team hung the STS-116 Mission Plaque (represented by Karl Silverman). Karl Silverman hung the plaque with Roger Smith and Jennifer Clevenger (FAO Team) holding the ladder.

STS-117	06/08/07	Hang the plaque - GNC/Mike Hamilton and the GNC team Hold the ladder - PDRS Team 4/Quincy Harp
STS-118	08/08/07	Christi Worstell, representing the EGIL Team, hung the plaque. Mike Scheib, representing the FAO Team, held the ladder
STS-120	10/23/07	GNC/David Weiler hung the Shuttle Mission plaque
STS-122	02/07/08	Lead FAO/Laura Hearon Hung the STS-122 Shuttle Plaque in the Shuttle FCR and Lead EVA Officer/Anna Jarvis hung the STS-122 Mission Plaque in the ISS FCR on behalf of the EVA Team.
STS-123	03/26/08	The EECOM team led by Rachel Hinterlang were selected to hang the plaque in the WFCR. The GC team led by Mike Marsh held the ladder.
STS-124	05/31/08	Terry Clancey and Gail Hansen/FAO team hung the plaque.
STS-126	11/14/08	ACO/Caroline Kostak hung the plaque for her team and the Cargo Transfer team lead by Jeremy Owen held the ladder.
STS-119	03/15/09	PROP/Lonnie Schmitt and FDO/David Mayhew held the ladder; GNC/Ramon Gonzalez hung the plaque
STS-125	05/11/09	The SM4 DX EVA team led by Tomas Gonzalez-Torres was selected to hang the plaque. The SM4 Payloads team led by Miker Meyer, the SM4 PDRS team led by Ian Mills, the SM4 FAO team led by Mike Scheib, and the SM4 DM Orbit Adjustment Assessment Team led by Bill Tracy were all selected to hold the ladder.
STS-127	07/15/09	FDO/Jamie Marshik hung the plaque; The ISS Ops Plan Team held the latter
STS-128	08/28/09	The RNDZ team lead by Jerry Yencharis, was selected to hang the plaque. The FAO team (lead by Jennifer Clevenger, Jason Mitz, Refugio Molina), the ACO team (lead by Greg Humble, Lisa Milam, Dustin Keiser) and the GC team (lead by Aaron Frith), were all selected to hold the ladder.
STS-129	11/16/09	ACO/Robert Napp hung the plaque; EECOM/Bobby Jarvis held the ladder
STS-130	02/08/10	PROP/Matt Anderson hung the plaque, and the GSFC Network Team, represented by Jim Bangerter and Melissa Blizzard as well as GC/Brian Jones held the ladder.
STS-131	04/05/10	FAO/Tobin Melroy hung the plaque; Gary Kilgo held the ladder
STS-132	05/14/10	INCO/Jonathan King hung the plaque
STS-133	02/24/11	FDO/Mark McDonald held the ladder; FAO/Lauren Knowlton and Ops Plan/Todd Eggleston hung the plaque
STS-134	05/16/11	PDRS/Zach Drewry hung the plaque. The STORRM team lead by Dave Harshmann, Roger Rojas, and Greg Humbler) and FAO Team (lead by Jennifer Clevenger, Jason Mintz, and Rocy Garcia) held the ladder
STS-135	08/18/11	INCO/Heidi Brewer hung the plaque and DPS/Amanda Coots held the ladder

APPENDIX B
Flight Control Room/Shuttle Flights Supported/Dates

Flight Control Room (FCR)	Flights Supported	Flight Dates
FCR-1 (second floor) Apollo Mission Control Skylab Mission Control Space Shuttle Mission Control	Apollo 1, Apollo 5 Apollo 7 Skylab 1-4 Apollo-Soyuz Test Project *[Shuttle era begins]* STS-1-4. STS-41-D. STS-51-A. STS- 51-B. STS-51-F. STS-61-A, 61-B, 61-C. *[post-Challenger era]* STS-26 STS-29, 30. STS-34. STS-32. STS-31, 41. STS-35, 37, 39, 40, 43, 48, 44, 42, 45, 49, 50, 46, 47, 52. *[end FCR-2 flight support]* STS-54, 56, 55, 57, 51, 58, 61, 60, 62, 59, 65, 64, 68, 66, 63, 67. STS-71 (first Mir). *[begin White FCR support]*	Jan. 1967, Jan. 1968 Oct. 1968 May 1973—Feb. 1974 TBS 1975 April 1981—July 82 Aug. 1984. Nov. 1984. April 1985. July 1985. Oct. 1985, Nov. 1985, Jan. 1986. Sept. 1988. March 1989, May 1989. Oct. 1989. Jan. 1990. April 1990, Oct. 1990. Dec. 1990, April 1991, April 1991, June 1991, Aug. 1991, Sept. 1991, Nov. 1991, Jan. 1992, March 1992, May 1992, June 1992, July 1992, Sept. 1992, Oct. 1992. Jan. 1993, April 1993, April 1993, June 1993, Sept. 1993, Oct. 1993, Dec. 1993, Feb. 1994, March 1994, April 1994, July 1994, Sept. 1994, Sept. 1994, Nov. 1994, Feb. 1995, March 1995, June 1995.
FCR-1 Reconfigured International Space Station Mission Control	Expedition 14. Expedition 15. Expedition 16 (CDR Peggy Whitson). Expedition 17. Expedition 18. Expedition 19. Expedition 20. Expedition 21. Expedition 22. Expedition 23. Expedition 24. Expedition 25. Expedition 26. Expedition 27. Expedition 28. *[end Shuttle Program]* Expeditions 29 to 62. SpaceX Crew Dragon Demo-2 Expedition 63 to present	Oct. 2006 to April 2007. April 2007 to Oct. 2007. Oct. 2007 to April 2008. April 2008 to Oct. 2008. Oct. 2008 to April 2009. March 2009. May 2009 to Oct. 2009 Sept. 2009 to Nov. 2009 Dec. 2009 to March 2010. April 2010 to June 2010. June 2010 to Sept. 2010. Oct. 2010 to Nov. 2010. Dec. 2010 to March 2011. April 2011 to May 2011. June 2011 to Sept. 2011. Nov. 2011 through May 2020 with crew rotations every 3 months via Soyuz TMA. May 2020 Exp. 63 up. May 2020 to present crew rotations via Russian and American spacecraft.
FCR-2 (3rd floor) Gemini Mission control Apollo Mission Control Shuttle Mission Control National Historic Monument	Gemini 4 to 12 Apollo 6. Apollo 8. Apollo 9 to 17 *[Shuttle era begins]* Shuttle ALT STS-5, 6, 7, 8, 9. STS-41-B, 41-C. STS-41-G. STS 51-C, 51-D. STS-51-G. STS 51-I, 51-J. STS 51-L Challenger lost	June 1965—Nov. 1966 April 1968 Dec. 1968 March 1969 through Dec. 1972 Aug.—Oct. 1977. Nov. 1982, April 1983, June 1983, Aug. 1983, Nov. 1983. Feb. 1984, April 1984. Oct. 1984. Jan. 1985, April 1985. June 1985. Aug. 1985, Oct. 1985. Jan. 1986.

	[post-Challenger era]	Dec. 1988
	STS-27.	Aug. 1989
	STS-28.	Nov. 1989
	STS-33.	Feb. 1990
	STS-36.	Nov. 1990
	STS-38.	Dec. 1992.
	STS-53.	Open to tours via Space Center Houston until 2018 when closed for renovation.
	[end FCR-2 flight support]	
	Apollo 11 Historic Landmark	July 2019 to present
White FCR (30 South) Space Shuttle Mission Control	STS-70, 69, 73, 74 (Mir), 72, 75, 76 (Mir). STS-77, 78, 79 (Mir), 80, 81 (Mir), 82, 83, 84 (Mir), 94, 85, 86 (Mir), 87, 89 (Mir), 90, 91 (last Mir), 95 (SpaceHab). STS-88 (first ISS), 96 (ISS), 93 (Chandra), 103 (Hubble), 99 (Radar), 101 (ISS), 106 (ISS), 92 (ISS), 97 (ISS), 98 (ISS), 102 (ISS), 100 (ISS), 104 (ISS), 105 (ISS), 108 (ISS), 109 (Hubble), 110 (ISS), 111 (ISS), 112 (ISS), 113 (ISS). STS-107 Columbia lost. STS-114, 121, 115, 116, 117, 118, 120, 122, 123, 124, 126, 119, 125 (last Hubble), 127, 128, 129, 130, 131, 132, 133, 134, 135 (last Shuttle). [end Shuttle era] Commercial Crew Program (Boeing CST-100)	July 1995 through March 1996 [These patches appear in the new FCR 1] May 1996 through November 1998. Dec. 1998 through Dec. 2002. Jan. 2003 All Shuttle flights from July 2005 through July 2011 went to the ISS except STS-125 which was the final Hubble repair mission. Unmanned test Dec. 2019
Blue FCR (30 South) Special Vehicle Operations Room (SVO) International Space Station Mission Control Orion Mission Control Commercial Crew Mission Control	STS-88, 96. STS-101, 106 (ISS 2A.2b), 92. Exp. 1 (up TMA) STS-97, 98, 102 (Exp. 1 down, Exp. 2 up). Exp. 2 STS-100, 104, 105 (Exp 2 down 3 up). Exp. 3 STS-108 (Exp 3 down Exp. 4 up). Exp. 4 STS-110, 111 (Exp 4 down 5 up). Exp. 5 STS-112, 113 (Exp 5 down 6 up). Exp. 6 [Columbia lost] Exp. 7. Exp. 8. Exp. 9. Exp. 10. Exp. 11, STS-114 Exp. 12 Exp. 13, STS-121 (Exp. 14 up). Exp. 14, STS-115 [begin ISS support from FCR-1] [end Shuttle era] EFT-1 Orion test flight Support for Commercial Crew	ISS assembly flights: Dec. 1998, May 1999. May 2000, Sept. 2000, Oct. 2000. Nov. 2000, start of 24-hour operations Nov. 2000, Feb. 2001, March 2001. March 2001 April 2001, July 2001, Aug. 2001. Aug. 2001 Dec. 2001 Dec. 2001 April 2002, June 2002. June 2002. Oct. 2002, Nov. 2002. Nov. 2002 to May 2003 (down TMA-1) Russian Soyuz for crew rotation. May 2003 to April 2003. Oct. 2003 to April 2004. April 2004 to October 2004. Oct. 2004 to April 2005. April 2005 to Oct. 2005, July 2005. Oct. 2005 to April 2006. March 2006 to Sept. 2006, July 2006. Sept. 2006 to April 2007, Sept. 2006. Dec. 2014 Unmanned test Dec. 2019

INDEX